Annals of Mathematics Studies

Number 152

Finite Structures
with Few Types

GREGORY CHERLIN AND
EHUD HRUSHOVSKI

PRINCETON UNIVERSITY PRESS

PRINCETON AND OXFORD

2003

The Annals of Mathematics Studies are edited by
John N. Mather and Elias M. Stein

Library of Congress Cataloging-in-Publication Data

Cherlin, Gregory L., 1948-
Finite structures with few types / by Gregory Cherlin and Ehud Hrushovski.
p. cm. – (Annals of mathematics studies ; no. 152)
Includes bibliographical references and index.
ISBN 0-691-11331-9 (alk. paper) – ISBN 0-691-11332-7 (pbk. : alk. paper)
1. Model theory. 2. Group theory. 3. Envelopes (Geometry)
I. Hrushovski, Ehud, 1959- II. Title. III. Series.
QA9.7 .C48 2003
511.3—dc21   200202980

The publisher would like to acknowledge the authors of this volume for
providing the camera-ready copy from which this book was printed

GC acknowledges support by NSF Grants DMS 8903006, 9121340, and 9501176.
Support of MSRI, Berkeley, and the Mathematics Institute of the Hebrew University,
Jerusalem is gratefully acknowledged. EH acknowledges research support by the
MSRI, Berkeley and by grants from the NSF and ISF.

Printed on acid-free paper. ∞

www.pupress.princeton.edu

10 9 8 7 6 5 4 3 2 1

Contents

1. Introduction   1

   1.1 The Subject   1
   1.2 Results   4

2. Basic Notions   11

   2.1 Finiteness Properties   11
      2.1.1 Quasifiniteness, weak or smooth approximability   11
      2.1.2 Geometries   13
      2.1.3 Coordinatization   16
   2.2 Rank   18
      2.2.1 The rank function   18
      2.2.2 Geometries   20
      2.2.3 A digression   22
   2.3 Imaginary Elements   23
   2.4 Orthogonality   31
   2.5 Canonical Projective Geometries   36

3. Smooth Approximability   40

   3.1 Envelopes   40
   3.2 Homogeneity   42
   3.3 Finite Structures   46
   3.4 Orthogonality Revisited   50
   3.5 Lie Coordinatization   54

4. Finiteness Theorems   63

   4.1 Geometrical Finiteness   63
   4.2 Sections   67
   4.3 Finite Language   71
   4.4 Quasifinite Axiomatizability   75
   4.5 Ziegler's Finiteness Conjecture   79

5. Geometric Stability Generalized   82

  5.1 Type amalgamation   82
  5.2 The sizes of envelopes   90
  5.3 Nonmultidimensional Expansions   94
  5.4 Canonical Bases   97
  5.5 Modularity   101
  5.6 Local Characterization of Modularity   104
  5.7 Reducts of Modular Structures   107

6. Definable Groups   110

  6.1 Generation and Stabilizers   110
  6.2 Modular Groups   114
  6.3 Duality   120
  6.4 Rank and Measure   124
  6.5 The Semi-Dual Cover   126
  6.6 The Finite Basis Property   134

7. Reducts   141

  7.1 Recognizing Geometries   141
  7.2 Forgetting Constants   149
  7.3 Degenerate Geometries   153
  7.4 Reducts with Groups   156
  7.5 Reducts   164

8. Effectivity   170

  8.1 The Homogeneous Case   170
  8.2 Effectivity   173
  8.3 Dimension Quantifiers   178
  8.4 Recapitulation and Further Remarks   183
       8.4.1 The role of finite simple groups   183

REFERENCES   187

INDEX   191

1

Introduction

1.1 THE SUBJECT

In the present monograph we develop a structure theory for a class of finite structures whose description lies on the border between model theory and group theory. Model theoretically, we study large finite structures for a fixed finite language, with a bounded number of 4-types. In group theoretic terms, we study all sufficiently large finite permutation groups which have a bounded number of orbits on 4-tuples and which are k-closed for a fixed value of k. The primitive case is analyzed in [KLM; cf. Mp2]. The treatment of the general case involves the application of model theoretic ideas along lines pioneered by Lachlan.

We show that such structures fall into finitely many classes naturally parametrized by "dimensions" in the sense of Lachlan, which approximate finitely many infinite limit structures (a version of Lachlan's theory of shrinking and stretching), and we prove uniform finite axiomatizability modulo appropriate axioms of infinity (quasifinite axiomatizability). We also deal with issues of effectivity. At our level of generality, the proofs involve the extension of the methods of stability theory—geometries, orthogonality, modularity, definable groups—to this somewhat unstable context. Our treatment is relatively self-contained, although knowledge of the model theoretic background provides considerable motivation for the results and their proofs. The reader who is more interested in the statement of precise results than in the model theoretic background will find them in the next section.

On the model theoretic side, this work has two sources. Lachlan worked out the theory originally in the context of stable structures which are homogeneous for a finite relational language [La], emphasizing the parametrization by numerical invariants. Zilber, on the other hand, investigated totally categorical structures and developed a theory of finite approximations called "envelopes," in his work on the problems of finite axiomatizability. The class of $\aleph_0$-categorical, $\aleph_0$-stable structures provides a broad model theoretic context to which both aspects of the theory are relevant. The theory was worked out at this level in [CHL], including the appropriate theory of envelopes. These were used in particular to show that the corresponding theories are not finitely axiomatizable, by Zilber's method. The basic tool used in [CHL], in accordance with Shelah's general approach to stability theory and geometrical refinements due to Zilber, was a "coordinatization" of an arbitrary structure in the class by a tree of standard coordinate geometries (affine or projective over finite fields, or degenerate. Other classical geometries involving quadratic forms were conspicuous only by their absence at this point.

The more delicate issue of finite axiomatizability modulo appropriate "axioms of infinity," which is closely connected with other finiteness problems as well as problems of effectivity, took some time to resolve. In [AZ1] Ahlbrandt and Ziegler isolated the relevant combinatorial property of the coordinatizing geometries, which we refer to here as "geometrical finiteness," and used it to prove quasifinite axiomatizability in the case of a single coordinatizing geometry. The case of $\aleph_0$-stable, $\aleph_0$-categorical structures in general was treated in [HrTC].

The class of *smoothly approximable* structures was introduced by Lachlan as a natural generalization of the class of $\aleph_0$-categorical $\aleph_0$-stable structures, in essence taking the theory of envelopes as a definition. Smoothly approximable structures are $\aleph_0$-categorical structures which can be well approximated by finite structures in a sense to be given precisely in §2.1. One of the achievements of the structure theory for $\aleph_0$-categorical $\aleph_0$-stable theories was the proof that they are smoothly approximable in Lachlan's sense. While this was useful model theoretically, Lachlan's point was that in dealing with the model theory of large finite structures, one should also look at the reverse direction, from smooth approximability to the structure theory. We show here, confirming this not very explicitly formulated conjecture of Lachlan, that the bulk of the structure theory applies to smoothly approximable structures, or even, as stated at the outset, to sufficiently large finite structures with a fixed finite language, having a bounded number of 4-types.

Lachlan's project was launched by Kantor, Liebeck, and Macpherson in [KLM] with the classification of the primitive smoothly approximable structures in terms of various more or less classical geometries (the least classical being the "quadratic" geometry in characteristic 2, described in §2.1.2). These turn up in projective, linear, and affine flavors, and in the affine case there are some additional nonprimitive structures that play no role in [KLM] but will be needed here ("affine duality," §2.3). Bearing in mind that any $\aleph_0$-categorical structure can be analyzed to some degree in terms of its primitive sections, the results of [KLM] furnish a rough coordinatization theorem for smoothly approximable structures. This must be massaged a bit to give the sort of coordinatization that has been exploited previously in an ω-stable context. We will refer to a structure as "Lie coordinatizable" if it is bi-interpretable with a structure which has a nice coordinatization of the type introduced below. Lie coordinatizability will prove to be equivalent to smooth approximability, in one direction largely because of [KLM], and in the other by the analog of Zilber's theory of envelopes in this context. One tends to work with Lie coordinatizability as the basic technical notion in the subject. The analysis in [KLM] was in fact carried out for primitive structures with a bound on the number of orbits on 5-tuples, and in [Mp2] it was indicated how the proof may be modified so as to work with a bound on 4-tuples. (Using only [KLM], we would also be forced to state everything done here with 5 in place of 4.)

In model theory, techniques for going from a good description of primitive pieces to meaningful statements about imprimitive structures generally fall under the heading of "geometrical stability theory," whose roots lie in early work of Zilber on $\aleph_1$-categorical theories, much developed subsequently. Though the present theory lies slightly outside stability theory (it can find a home in the more recent developments relating to simple theories), geometrical stability theory provided a very useful template [Bu, PiGS].

Before entering into greater detail regarding the present work, we make some comments on the Galois correspondence between structures and permutation groups implicit in the above, and on its limitations.

Let X be a finite set. There is then a Galois correspondence between subgroups of the symmetric group $Sym(X)$ on X, and model theoretic structures with universe X, associating to a permutation group the invariant relations, and to a structure its automorphism group. This correspondence extends to $\aleph_0$-categorical structures ([AZ1, Introduction], [CaO]).

When we consider infinite families of finite structures in general, or a passage to an infinite limit, this correspondence is not well behaved. For instance, the automorphism group of a large finite random graph of order n (with constant and nontrivial edge probability) is trivial with probability approaching 1 as n goes to infinity, while the natural model theoretic limit is the random countable graph, which has many automorphisms.

It was shown in [CHL], building on work of Zilber for totally categorical structures, that structures which are both $\aleph_0$-categorical and $\aleph_0$-stable can be approximated by finite structures simultaneously in both categories. Lachlan emphasized the importance of this property, which will be defined precisely in §2.1, and proposed that the class of structures with this property, the *smoothly approximable structures*, should be amenable to a strong structure theory, appropriately generalizing [CHL]. Moreover, Lachlan suggested that the direction of the analysis can be reversed, from the finite to the infinite: one could classify the large finite structures that appear to be "smooth approximations" to an infinite limit, or in other words, classify the families of finite structures which appear to be Cauchy sequences both as structures and as permutation groups. This line of thought was suggested by Lachlan's work on stable finitely homogeneous structures [La], much of which predates the work in [CHL], and provided an additional ideological framework for that paper.

In the context of stable finitely homogeneous structures this analysis in terms of families parametrized by dimensions was carried out in [KL] (cf. [CL, La]), but was not known to go through even in the totally categorical case. Harrington pointed out that this reversal would follow immediately from compactness if one were able to work systematically within an elementary framework [Ha]. This idea is implemented here: we will replace the original class of "smoothly approximable structures" by an elementary class, a priori larger. Part of our effort then goes into developing the structure theory for the ostensibly broader class.

From the point of view of permutation group theory, it is natural to begin the analysis with the case of finite primitive structures. This was carried out using group theoretic methods in [KLM], and we rely on that analysis. However, there are model theoretic issues which are not immediately resolved by such a classification, even for primitive structures. For instance, if some finite graphs G_n are assumed to be primitive, and to have a uniformly bounded number of 4-types, our theory shows that an ultraproduct G^* of the G_n is bi-interpretable with a Grassmannian structure, which does not appear to follow from [KLM] by direct considerations. The point here is that if G_n is "the same as" a Grassmannian structure in the category of permutation groups, then it is bi-interpretable with such a structure on the model theoretic side. To deal with families, one must deal (at least implicitly) with the uniformity of such interpretations; see §8.3, and the sections on reducts. It is noteworthy that our proof in this case actually passes through the theory for imprimitive structures: any nonuniform interpretation of a Grassmannian structure on G_n gives rise to a certain structure on G^*, a reduct of the structure which would be obtained from a uniform interpretation, and one argues that finite approximations (on the model theoretic side) to G^* would have too many automorphisms. In other words, we can obtain results on uniformity (and hence effectivity) by ensuring that the class for which we have a structure theory is closed under reducts. This turns out to be a very delicate point, and perhaps the connection with effectivity explains why it should be delicate.

1.2 RESULTS

A rapid but thorough summary of this theory was sketched in [HrBa], with occasional inaccuracies. For ease of reference we now repeat the main results of the theory as presented there, making use of a considerable amount of specialized terminology which will be reintroduced in the present work. The various finiteness conditions referred to are all given in Definition 2.1.1.

Theorem 1 (Structure Theory)
   Let M be a Lie coordinatizable structure. Then M can be presented in a finite language. Assuming M is so presented, there are finitely many definable dimension invariants for M which are infinite, up to equivalence of

such invariants. If C is a set of representatives for such definable dimension invariants, then there is a sentence $\varphi = \varphi_{\mathcal{M}}$ with the following properties:

1. *Every model of φ in which the definable dimension invariants of C are well-defined is determined up to isomorphism by these invariants.*
2. *Any sufficiently large reasonable sequence of dimension invariants is realized by some model of φ.*
3. *The models of φ for which the definable dimension invariants of C are well-defined embed homogeneously into $\mathcal{M}$ and these embeddings are unique up to an automorphism of $\mathcal{M}$.*

There are a considerable number of terms occurring here which will be defined later. Readers familiar with "shrinking" and "stretching" in the sense of Lachlan should recognize the situation. Definable dimension invariants are simply the dimensions of coordinatizing geometries which occur in families of geometries of constant dimension; when the appropriate dimensions are not constant within each family, the corresponding invariants are no longer well-defined. A dimension invariant is reasonable if its parity is compatible with the type of the geometry under consideration; in particular, infinite values are always reasonable.

The statements of the next two theorems are slight deformations of the versions given in [HrBa]. We include more clauses here, and we use definitions which vary slightly from those used in [HrBa].

Theorem 2 (Characterizations)

The following conditions on a model $\mathcal{M}$ are equivalent:

1. *$\mathcal{M}$ is smoothly approximable.*
2. *$\mathcal{M}$ is weakly approximable.*
3. *$\mathcal{M}$ is strongly quasifinite.*
4. *$\mathcal{M}$ is strongly 4-quasifinite.*
5. *$\mathcal{M}$ is Lie coordinatizable.*
6. *The theory of $\mathcal{M}$ has a model $\mathcal{M}^*$ in a nonstandard universe whose size is an infinite nonstandard integer, and for which the number of internal n-types $s_n^*(\mathcal{M}^*)$ satisfies*

$$s_n^*(\mathcal{M}^*) \leq c^{n^2}$$

for some finite c, and in which internal n-types and n-types coincide. (Here n varies over standard natural numbers.)

The class characterized above is not closed under reducts. For the closure under reducts we have:

Theorem 3 (Reducts)

The following conditions on a model $\mathcal{M}$ are equivalent:

1. $\mathcal{M}$ has a smoothly approximable expansion.
2. $\mathcal{M}$ has a weakly approximable expansion.
3. $\mathcal{M}$ is quasifinite.
4. $\mathcal{M}$ is 4-quasifinite.
5. $\mathcal{M}$ is weakly Lie coordinatizable
6. The theory of $\mathcal{M}$ has a model $\mathcal{M}^*$ in a nonstandard universe whose size is an infinite nonstandard integer, and for which the number of internal n-types $s_n^*(\mathcal{M}^*)$ satisfies:

$$s_n^*(\mathcal{M}^*) \leq c^{n^2}$$

for some finite c. (Here n varies over standard natural numbers.)

On the other hand, once the class is closed under reducts it is closed under interpretation, hence:

Theorem 4 (Interpretations)
   The closure of the class of Lie coordinatizable structures under interpretation is the class of weakly Lie coordinatizable structures.

An earlier claim that the class of Lie coordinatizable structures is closed under interpretations was refuted by an example of David Evans which will be given below.

Theorem 5 (Decidability)
   For any k and any finite language, the theory of finite structures with at most k 4-types is decidable, uniformly in k. The same applies in an extended language with dimension comparison quantifiers and Witt defect quantifiers. Thus one can decide effectively whether a sentence in such a language has a finite model with a given number of 4-types.

This is a distant relation of a family of theorems in permutation group theory giving explicit classifications of primitive permutation groups with very few 2-types. Dimension comparison quantifiers do not allow us to quantify over the dimensions of spaces, but they allow us to compare the dimensions of any two geometries. Witt defect quantifiers are more technical (§2.1, Definition 2.1.1).

Theorem 6 (Finite structures)
   Let L be a finite language and k a natural number. Then the class of finite L-structures having at most k 4-types can be divided into families $\mathcal{F}_1, \dots, \mathcal{F}_n$ for some effectively computable n such that

1. Each family $\mathcal{F}_i$ is finitely axiomatizable in a language with dimension comparison and Witt defect quantifiers.
2. Each family $\mathcal{F}_i$ is associated with a single countable Lie coordinatizable structure $\mathcal{M}_i$. The family $\mathcal{F}_i$ is the class of "envelopes" of $\mathcal{M}_i$, which are the structures described in Theorem 1, parametrized by freely vary-

ing definable dimension invariants (above a certain minimal bound, with appropriate parity constraints).

3. *For $\mathcal{M}$, $\mathcal{N}$ in $\mathcal{F}_i$, if the dimension invariants satisfy $d(\mathcal{M}) \leq d(\mathcal{N})$ then there is a homogeneous embedding of $\mathcal{M}$ in $\mathcal{N}$, unique up to an automorphism of $\mathcal{N}$.*

4. *Membership in each of the families $\mathcal{F}_i$ (and in particular, in their union) can be determined in polynomial time, and the dimension invariants can be computed in polynomial time. Thus the isomorphism problem in the class of finite structures with a bounded number of types can be solved in polynomial time.*

5. *The cardinality of an envelope of dimension d is an exponential polynomial in d; specifically, a polynomial in exponentials of the entries of d (with bases roughly the sizes of the base fields involved). The structure $N_i(d)$ which is the member of $\mathcal{F}_i$ of specified dimensions d can be constructed in time which is polynomial in its cardinality.*

Theorem 7 (Model Theoretic Analysis)

The weakly Lie coordinatizable structures $\mathcal{M}$ are characterized by the following nine model theoretic properties:

LC1. *$\aleph_0$-categoricity.*

LC2. *Pseudofiniteness.*

LC3. *Finite rank.*

LC4. *Independent type amalgamation.*

LC5. *Modularity in $\mathcal{M}^{\mathrm{eq}}$.*

LC6. *The finite basis property in groups.*

LC7. *General position of large 0-definable sets.*

LC8. *$\mathcal{M}$ does not interpret the generic bipartite graph.*

LC9. *For every vector space V interpreted in $\mathcal{M}$, the definable dual V^* (the set of all definable linear maps on V) is interpreted in $\mathcal{M}$.*

Some of these notions were first introduced in [HrBa], sometimes using different terminology. In particular, the rank function is not a standard rank function, the finite basis property in groups (or "linearity") reduces to local modularity in the stable case, and the general position (or "rank/measure") property is an additional group theoretic property that arises in the unstable case, when groups tend to have many definable subgroups of finite index. The eighth condition is peculiarly different from the ninth. This is a corrected version of Theorem 6 of [HrBa].

David Evans made several contributions to the theory given here, notably the observation that the orientation of quadratic geometries is essential, and bears on the problem of reducts. The detection of all such points is critical. Evans also gave a treatment of weak elimination of imaginaries in linear geometries, in [EvSI].

We will say a few words about the development of this material, using technical notions explained fully in the text. The first author on reading [KLM] understood that one could extract stably embedded geometries from the analysis of primitive smoothly approximable structures given there, and that the group theory gives a decent orthogonality theory (but the orthogonality theory given here will be based more on geometry than on group theory). These ingredients seemed at first to be enough to reproduce the Ahlbrandt–Ziegler analysis, after the routine verification that the necessary geometrical finiteness principle follows from Higman's lemma; all of this follows the lead of [AZ1], along the lines developed in [HrTC]. An attempt to implement this strategy failed, in part because at this stage there was no hint of "affine duality."

The second author then produced affine duality and gave a complete proof of quasifinite axiomatizability, introducing some further modifications of the basic strategy, notably canonical projectives and a closer analysis of the affine case. The theme in all of this is that one should worry even more about the interactions of affine geometries than one does in the stable case. This can perhaps be explained by the following heuristic. Only the projective geometries are actually coordinatizing geometries; the linear and affine geometries are introduced to analyze definable group structures, in keeping with the general philosophy that structures are built from basic 1-dimensional pieces, algebraic closure, and definable groups. Here higher dimensional groups are not needed largely because of the analog of 1-basedness, referred to below as the finite basis property. The developments that go beyond what is needed for quasifinite axiomatizability are all due to the second author. The extension of a considerable body of geometric stability theory to this context is essential to further developements. The high points of these developments, as far as applications are concerned, are the analysis of *reducts* and its applications to issues of *effectivity*. It may be noted also that the remarkable quadratic geometries have been known for some time, and play an essential role in [KLM], in particular. In our view they add considerably to the appeal of the theory.

The treatment of reducts requires a considerably more elaborate transference of techniques of stability theory to this unstable setting than would be required for the quasifinite axiomatizability alone. This would not be indispensable for the treatment of structures already equipped with a Lie coordinatization; but to apply these results to classes which are closed under interpretation requires the ability to recognize an appropriate coordinatization, starting from global properties of the structure; thus one must find the model theoretic content of the property of coordinatizability by the geometries on hand.

Our subject has also been illuminated by recent developments in connection with Shelah's "simple theories," and is likely to be further illuminated by that theory.

Various versions of this material, less fully worked out, have been in circulation for a considerable period of time (beginning with notes written in Spring

1990) and have motivated some of the work in simple theories. In particular, versions of sections 5.1 [KiP], 5.4, and 6.1 [PiGr] have been obtained in that very general context; all of this rests on the theoretical foundation provided by the original paper of Shelah [ShS] and subsequent work by Kim [Ki].

Some comments on the relationship of this theory to Shelah's "simple theories" are in order. Evidently a central preoccupation of the present work is the extension of methods of stability theory to an unstable context. Stability theory is a multilayered edifice. The first layer consists of a theory of rank and the related combinatorial behavior of definable sets. The next layer includes the theory of orthogonality, regular types, and modularity, and was initially believed to be entirely dependent on the foundational layer in its precise form. One of the key conclusions of the present work is that is possible to recover the second "geometric model theory" layer over an unstable base. Because we have $\aleph_0$-categoricity and finiteness of the rank, our basic rank theory becomes as simple as possible; nonetheless, almost all of the "second-level" phenomena connected with simplicity appear in our context with their full complexity—the main exception being the Lascar group. It was perhaps this combination of circumstances that facilitated a very successful generalization of the "geometric theory" to the simple context, once the first layer was brought into an adequate state by Kim's thesis [KiTh].

As far as the present work is concerned, the development of a sufficiently general theory was often due to necessity rather than insight. For example, if we—or the creator of the finite simple groups—had been able to exclude from consideration the orthogonal geometries in characteristic 2, we would have had a considerably simpler theory of generics in groups, with $Stab = Stab_o$ (cf. §6.1, Definition 6.1.9, and the Example following). Such a simplified theory would have been much less readily generalizable to the simple context; in addition, under the same hypothesis, this simplified theory would have largely obviated the need for the theory of the semi-dual cover.

A number of features of the theory exposed here have been generalized with gratifying success to the context of simple theories, but some have not. On the positive side, one has first of all the theorem which we originally called *the independence theorem*. This name has become standard in the literature, although in the present manuscript it was eventually renamed "the type amalgamation property." In any case this is still a misnomer, as this amalgamation involves a triple over a base rather than a pair. Compare the following "homological" description. Let $I(n)$ be the space of n-types, over some fixed base, of *independent* n-tuples (whose elements are themselves finite sequences of elements). We have "projection" maps $\pi_i : I(n) \to I(n-1)$ obtained by deletion of one coordinate. The uniqueness of forking in stability theory is the statement that the induced map $I(2) \to I(1)^2$ is *injective*. We replace this by an *exactness* property, characterizing the image of $I(3)$ in $I(2)^3$ by minimal coherence conditions.

The first proof found for this theorem consisted of inspection in the 1-dimensional case, followed by an induction on rank. In the course of related work, an abstract proof was found, assuming finite simplicity rank and definability of the rank. This proof was later generalized by Kim and Pillay, and together with their realization of the relevance of the Lascar group, it became the central pillar of simplicity theory. In §5.1 we retain the original clumsy inductive proof. This may be of use in situations where simplicity is not known in advance.

The main point in any case is not the proof of this theorem but the realization that the uniqueness of nonforking extensions, which seemed characteristic of stability theory and essential to its fabric, can be replaced "densely often" with an appropriate *existential* statement.

The definition of *modularity* could largely be taken over from the stable case. A new idea was required (cf. §5.4) to produce enough geometric imaginaries for proof of the local–global principle; this idea survives in the contemporary treatment of canonical bases in simple theories. The consequences of modularity for groups are not as decisive in general as in the stable case, even generically, so we had to consider stronger variants. The recognition theorems in rank one which use these properties serve to situate the basic geometries model theoretically to a degree. One would like to see these theorems generalized, as Zilber's characterizations of modular groups were extended from the totally categorical to the strongly minimal case.

The strong presence of duality is also a new feature as far as the model theory is concerned. Initially it arose as a particular instance of instability, which we sought to circumscribe and neutralize as much as possible. At the outset duals must be recognized in order to render the basic geometries stably embedded; the dual space of a finite vector space is also a prime example of a nonuniform interpretation. Eventually duality also emerged as a positive tool, useful for certain purposes even in contexts where stability is initially assumed: see §6.5, on the semi-dual cover, and also the treatment of second-order quantifiers in Chapter 8, dealing with effectivity. It seems possible that linear duality, like modularity, has some significance in general model theoretic frameworks, but at this time our situation remains isolated, awaiting further illumination.

The proof of Theorem 2 will be largely complete by the end of §3.5 (see the discussion in §3.5 for more on this). The final section (§8.4) contains some retrospective remarks on the structure of our development.

Various versions of this paper have benefited from remarks by a variety of model theorists. We thank particularly Ambar Chowdhury, David Evans, Bradd Hart, Dugald Macpherson, Anand Pillay, and Frank Wagner for their remarks. We thank Virginia Dunn, Amélie Cherlin, and Jakob Kellner for various forms of editorial and technical assistance. The first author also thanks Amaal for diverting correspondence during the preparation of the final version.

2

Basic Notions

2.1 FINITENESS PROPERTIES

We discuss at length the various finiteness properties to be considered here.

We will make use of nonstandard terminology as a convenient way of dealing with "large" integers; see [FJ, Chapter 13] (in particular, the examples treated therein, in §13.5) for a full presentation of this method. The method is based on the idea of replacing the standard model of set theory in which one normally works by a proper elementary extension, the "enlargement," in which there are "new" (hence, infinite) integers. Since the extension is elementary, all notions of set theory continue to have meaning, and (more or less) their usual properties. In particular, for any set S occurring in the enlargement, there is an associated collection of "all" subsets of S in the sense of the enlargement; this will not actually contain all subsets of S in general, and those which are in fact present in the enlargement are called "internal" (the others could be called "external," but we do not use them). The word "internal" is used in other related ways: we may call an internal set which is finite in the sense of the enlargement either "internally finite," or "nonstandardly finite." A subset of an internally finite set need not be internal, but if it is, it will be internally finite. Again, we refer to the presentation by Fried and Jarden [FJ] for the essential foundational material.

2.1.1 Quasifiniteness, weak or smooth approximability

Definition 2.1.1. *Let $\mathcal{M}$ be a structure.*

1. *$\mathcal{M}$ is $\aleph_0$-categorical, or* oligomorphic, *if for each n $\mathcal{M}$ has finitely many n-types.*

2. *$\mathcal{M}$ is* pseudofinite *if it is a model of the theory of all finite structures (in the same language).*

3. *$\mathcal{M}$ is k-quasifinite if in a nonstandard extension of the set theoretical universe it is elementarily equivalent to an internally finite model with finitely many internal k-types.*

4. *$\mathcal{M}$ is* quasifinite *if in a nonstandard extension of the set theoretical universe it is elementarily equivalent (in the original language L) to an internally finite L^*-structure with a finite number of internal k-types, for all k.*

5. *A finite substructure N of M is k-homogeneous in M if all 0-definable relations on M induce 0-definable relations on N, and for every pair of k-tuples $\mathbf{a}$, $\mathbf{b}$ in N, $\mathbf{a}$ and $\mathbf{b}$ have the same type in N if and only if they have the same type in M.*

6. *A structure M is* weakly approximable *by finite structures if it is $\aleph_0$-categorical, and every finite subset X of M is contained in a finite substructure N which is $|X|$-homogeneous in M.*

7. *A structure M is* smoothly approximable *by finite structures if it is $\aleph_0$-categorical, and every finite subset X of M is contained in a finite substructure N which is $|N|$-homogeneous in M.*

8. *M is* strongly k-quasifinite *if in a nonstandard extension of the set theoretical universe it is elementarily equivalent to an internally finite model with finitely many internal k-types, which coincide with the k-types.*

9. *M is* strongly quasifinite *if in a nonstandard extension of the set theoretical universe it is elementarily equivalent (in the original language L) to an internally finite L^*-structure with a finite number of internal k-types, which coincide with the k-types, for all k.*

Remarks 2.1.2

We use freely the usual characterizations of $\aleph_0$-categoricity. Pseudofiniteness is also commonly referred to as the *finite model property*. Quasifiniteness strengthens pseudofiniteness (which is perhaps etymologically incorrect), as one sees by expressing pseudofiniteness in nonstandard terms. It also implies $\aleph_0$-categoricity, since the condition on internal k-types is equivalent to a similar condition on internal formulas with k free variables, and this includes the standard formulas. Decoding the nonstandard formulation yields:

3'. A structure M is k-quasifinite if and only if there is a finite number N such that for an arbitrary sentence φ true in M, there is a finite structure N satisfying φ in which there are at most N formulas in k free variables.

4'. A structure M is quasifinite if and only if there is a function $\nu : \mathbb{N} \to \mathbb{N}$ such that for any n and an arbitrary sentence φ true in M, there is a finite structure N satisfying φ in which there are at most $\nu(k)$ formulas in k free variables for $k \leq n$.

For strong quasifiniteness one specifies the formulas rather than the number of formulas.

Note that a weakly approximable structure M is strongly quasifinite, using the formulas which define k-types in a finite k-homogeneous substructure.

One gets an equivalent notion by bounding types rather than formulas, or equivalently, by bounding the number of orbits of the automorphism group of N on k-tuples. This concept would seem to be the most natural one from a purely permutation group theoretic standpoint. The definition of (strong) quasifiniteness implies (strong) k-quasifiniteness for all k, but the converse is not immediate. As noted in §1, Theorem 2, we will show that (strong) 4-quasifiniteness (or using only [KLM]: (strong) 5-quasifiniteness) already implies (strong) quasifiniteness, so in particular this converse does hold. One might have the impression that $\aleph_0$-categorical pseudofinite structures are strongly quasifinite in general, but this is very far from the case. The generic graph seems to be the canonical counterexample; it is not quasifinite. The point is that while one might reasonably expect the property: "*every formula in k variables is equivalent to one in a specified finite set of formulas in k variables*" to be first order, it is not, in general.

As defined here all of these notions are invariant under elementary equivalence. When $\mathcal{M}$ is countable, weak and smooth approximability can be expressed somewhat more concretely in the form that $\mathcal{M}$ is a union of a countable chain of finite substructures $\mathcal{M}_i$ such that $\mathcal{M}_i$ is i-homogeneous (in the weak case), or $|\mathcal{M}_i|$-homogeneous (in the smooth case), respectively.

Digression 2.1.3

It is generally assumed that there is *not* going to be a coherent structure theory for $\aleph_0$-*categorical pseudofinite structures in a finite language*, though there is no solid evidence for this. One complication is that it seems to be quite hard in practice to determine whether a given finitely homogeneous structure is pseudofinite. For finitely homogeneous structures, pseudofiniteness holds in the stable case [La], fails in cases involving nondegenerate partial orders, and is obscure in most other cases, apart from those amenable to probabilistic analysis. The test case would be whether the generic triangle-free graph is pseudofinite.

2.1.2 Geometries

We have described most of the finiteness notions occurring in the statement of Theorem 2, with the exception of the technical notion of coordinatizability by Lie geometries. This notion in its most useful form involves some detailed properties of specific geometries. The relevant collection of geometries was given almost completely in [KLM], with the exception of what we call *affine duality*, which was not needed there. In addition a certain coordinatization theorem was proved there, which requires a further laying on of hands before it acquires the form most useful for a model theoretic analysis. We will now present the relevant geometries, which give first in their linear forms, and then in projective and affine versions. It should be borne in mind that geometries are understood to be structures in the model theoretic sense, and not simply lattices or combinatorial geometries.

Definition 2.1.4. *A* weak linear geometry *is a structure of one of the following six types, and a* linear geometry *is an expansion of a weak one by the introduction of a set of algebraic constants in* $\mathcal{M}^{\mathrm{eq}}$.

1. *A degenerate space: a pure set, with equality alone.*
   We tend to ignore this case, as our claims are trivial in this context. One may perhaps pretend that it is a vector space over a field of order 1, and that linear dependence over a set is membership; in this case it equals its projectivization and has no affine version.

2. *A pure vector space: (V, K), with K a finite field and V a K-vector space, with the usual algebraic structure.*
   Scalar multiplication is treated as a map from $K \times V$ to V rather than as a set of unary operators. This allows the Galois group of K to act on the structure.

3. *A polar space: $(V \cup W, K, L; \beta)$, where K is a finite field, L a K-line (1-dimensional K-space), V and W are K-spaces, and there is a nondegenerate bilinear pairing $\beta : V \times W \to L$.*
   We write $V \cup W$ rather than V, W because we treat $V \cup W$ as a set on which there is an equivalence relation with two classes, thereby preserving the symmetry between V and W. In particular, the domain of β is actually $(V \times W) \cup (W \times V)$, and β is symmetric.

4. *An inner product space: (V, K, L, β) where K is a finite field, L a K-line, $\beta : V \times V \to L$ a nondegenerate sesquilinear form with respect to a fixed automorphism σ with $\sigma^2 = 1$, and either σ is trivial and β is symplectic, or σ is nontrivial and β is hermitian with respect to σ.*
   (The symmetric case is included in the following class.)

5. *An orthogonal space: (V, K, L, q) where K is a finite field, L a K-line, and q a quadratic form on V with values in L, whose associated bilinear form is nondegenerate.*
   This point of view allows a treatment independent of the characteristic.

6. *A quadratic geometry: $(V, Q, K; \beta_V, +_Q, -_Q, \beta_Q, \omega)$, where K is a finite field of characteristic 2, V is a K-vector space, β_V is a nondegenerate symplectic bilinear form on V, Q is a set of quadratic forms q on V for which the associated bilinear form $q(v + w) + q(v) + q(w)$ is β_V, chosen so that V acts regularly on Q by translation, with $\beta_Q, +_Q, -_Q$ giving the interaction between Q and V, and ω specifying the Witt defect [CoAt], which is fairly obscure in the infinite dimensional case.*
   There is, evidently, a considerable amount to be elucidated here.
   In the first place, there are always quadratic forms q for which the associated bilinear form $q(v + w) + q(v) + q(w)$ is the given symplectic form β_V, and any two of them differ by a quadratic form which is additive; this is just the square of a K-linear map. The full linear dual V^* acts regularly by $q \mapsto q + \lambda^2$ ($q \in Q$, $\lambda \in V^*$) on this set of quadratic forms, and via the identification of V with a subspace of V^*, coming from the given symplectic inner product β_V, we get a semiregular action of V on this space of quadratic forms. Q will be one of the V-orbits. We take $\beta_Q : Q \times V \to K$ to be the evaluation map

$\beta_Q(q, v) = q(v)$, while $+_Q : V \times Q \to Q$ is the regular action of V on Q and $-_Q : Q \times Q \to V$ the corresponding "subtraction" map; both of these are definable from β_Q, e.g.: $v +_Q q = q + \lambda_v^2$ where λ_v is the linear form $\beta_V(v, \cdot)$. The map $\omega(q)$ is not definable from β_Q. In the finite $(2n)$ dimensional case it will give the Witt defect $\pm$ of q, which is the difference between n and the dimension of a maximal totally q-isotropic subspace; this is either 0 or 1. In the infinite dimensional case we require a different description. For $q_1, q_2 \in Q$, $\sqrt{q_1 + q_2}$ is a linear function of the form λ_v for a unique $v \in V$. Identifying v and λ_v, we may write $q(\sqrt{q_1 + q_2}) \in K$; furthermore, we find $q_1(\sqrt{q_1 + q_2}) = q_2(\sqrt{q_1 + q_2})$, which translates to $(v, v) = 0$. We will write $[q_1, q_2]$ for $q_1(\sqrt{q_1 + q_2})$. For $q_1, q_2, q_3 \in Q$ if $v = \sqrt{q_1 + q_2}$, $w = \sqrt{q_1 + q_3}$, and $\alpha = (v, w)$ we find $[q_1, q_2] + [q_1, q_3] + [q_2, q_3] = \tau(\alpha)$ with $\tau(x) = x^2 + x$ the Artin–Schreier polynomial. Hence the relation $[q_1, q_2] \in \tau[K]$ is an equivalence relation with two classes. ω has the effect of naming these classes as unary predicates. We will construe ω as a function from Q to $\{0, 1\} \subseteq K$. In particular, the Witt defect is taken modulo 2, which is quite convenient since it is then additive with respect to orthogonal sums.

Remarks 2.1.5

1. In the case of polar geometries we may write $W = V^*$ and $V = W^*$, informally, but as we are dealing with infinite dimensional spaces this should not be taken too literally. One can give this a precise sense if one associates with each of V and W the corresponding weak topology on its companion, making each the continuous dual of the other.

2. We use K-lines L rather than K itself in order to allow certain permutations of the language as automorphisms. The point is that if f is a bilinear form or a quadratic form and α is a scalar, then αf is another form of the same type with the same automorphism group. It will be convenient to view two structures with the same underlying set whose forms differ by a scalar as isomorphic. If α is a square they are isomorphic via multiplication by $\sqrt{\alpha}$, but in our formalism the identity map on the space extends to an isomorphism by allowing α to act on L. The same effect would be achieved by replacing the L-valued form f by the set of K-valued forms $\{\alpha f : \alpha \in K^\times\}$ and allowing scalars to act on the set of forms.

3. We can view a geometry as having as its underlying set a vector space in most cases, or a pair of spaces in duality in the polar case, or the set (V, Q) in the quadratic case, with the additional structure encoded in $\mathcal{M}^{\mathrm{eq}}$.

Definition 2.1.6

1. *An* unoriented weak *linear geometry is defined as one of the six types of geometry listed above, with the proviso that in the sixth case we omit the Witt defect function* ω.

2. *A* basic linear *geometry is a linear geometry in which the elements of* K

*and L are named, and in the polar case the two spaces V and W are named
(or, equivalently, treated as unary predicates).*

Definition 2.1.7
1. *A* projective geometry *is the structure obtained from a linear geometry
by factoring out the equivalence relation defined by* $acl(x) = acl(y)$, *with
algebraic closure understood in the model theoretic sense.*
2. *A* semiprojective geometry *is the structure obtained from a basic linear
geometry by factoring out the relation* $x^Z = y^Z$, *where Z is the center
of the automorphism group, that is, the set of scalars respecting any additional structure present. For example, in the symplectic case, the symplectic
scalars are* ± 1.

After we check quantifier elimination in basic linear geometries, it will be
clear that this algebraic closure operation is just linear span (in the sense appropriate to each case) and that our projective geometries are indeed projective
geometries in the nonquadratic case; in the polar case we will have two projective spaces (PV, PV^*) with a notion of perpendicularity.

Definition 2.1.8. *If V is a definable vector space and A is a definable set,
then A is an* affine V-space *if V acts definably and regularly on A. If J is
a linear geometry and V is its underlying vector space (or one of the two
underlying vector spaces in the polar case) then an* affine geometry (J, A)
*is a structure in which J carries its given structure and A carries the action
of V, with no further structure.*

We will deal subsequently with the model theoretic properties of linear,
affine, and projective geometries, but first we will deal with the notion of coordinatization that enters into the statement of Theorem 2 from Chapter 1

2.1.3 Coordinatization

Definition 2.1.9. *Let* $\mathcal{M} \subseteq \mathcal{N}$ *be structures with* $\mathcal{M}$ *definable in* $\mathcal{N}$, *and let*
$a \in \mathcal{N}^{eq}$ *represent the set* $\mathcal{M}$ *(its so-called canonical parameter).*
1. $\mathcal{M}$ *is* canonically embedded *in* $\mathcal{N}$ *if the 0-definable relations of* $\mathcal{M}$ *are
the relations on* $\mathcal{M}$ *which are a-definable in the sense of* $\mathcal{N}$.
2. $\mathcal{M}$ *is* stably embedded *in* $\mathcal{N}$ *if every* $\mathcal{N}$-definable relation on $\mathcal{M}$ is $\mathcal{M}$-
definable, uniformly. The uniformity can be expressed either by requiring
that the form of the definition over* $\mathcal{M}$ *be determined by the form of the
definition over* $\mathcal{N}$, *or by requiring that the same condition apply to all elementary extensions of the pair* $(\mathcal{M}, \mathcal{N})$.
3. $\mathcal{M}$ *is* fully embedded *in* $\mathcal{N}$ *if it is both canonically and stably embedded
in* $\mathcal{N}$.

Definition 2.1.10. *A structure $\mathcal{M}$ is* coordinatized by Lie geometries *if it carries a tree structure of finite height with a unique, 0-definable root, such that the following coordinatization and orientation properties hold.*

1. *(Coordinatization) For each $a \in \mathcal{M}$ above the root either a is algebraic over its immediate predecessor in the tree ordering, or there exists $b < a$ and a b-definable projective geometry J_b fully embedded in $\mathcal{M}$ such that either*

    (i) *$a \in J_b$; or*

    (ii) *there is c in $\mathcal{M}$ with $b < c < a$, and a c-definable affine or quadratic geometry (J_c, A_c) with vector part J_c, such that $a \in A_c$ and the projectivization of J_c is J_b. (Note that the projectivization of a symplectic geometry in characteristic 2 may have both quadratic and affine geometries attached to it in this way.)*

2. *(Orientation) If $a, b \in \mathcal{M}$ have the same type and are associated with coordinatizing quadratic geometries J_a, J_b in $\mathcal{M}$, then there is no definable orientation-reversing isomorphism of J_a and J_b as unoriented weak linear geometries; in other words, if a definable map between them preserves everything other than ω, then it also preserves ω.*

Example 2.1.11

Let A be the infinite direct sum of copies of $(\mathbb{Z}/p^2\mathbb{Z})$ with p a fixed prime. One coordinatizes this by placing 0 at the root, as a finite set, then putting the projectivization of

$$A[p] = \{a \in A : pa = 0\}$$

above it, and $A[p] \backslash \{0\}$ itself above that (covering each projective point by the corresponding finite set of points above it); finally, one adds $A \backslash A[p]$; above each $a \in A[p] \backslash \{0\}$ one has the affine space $A_a = \{x \in A : px = a\}$. This gives a tree of height 4, with layers of the form: finite, projective, finite, affine, respectively.

We also use the briefer expression *Lie coordinatized* with the same meaning. However, we make a rather sharp distinction between the existence of a coordinatization, as defined above, and coordinatizability in the following more general sense.

Definition 2.1.12. *A structure $\mathcal{M}$ is* Lie coordinatizable *if it is bi-interpretable with a structure having finitely many 1-types which is coordinatized by Lie geometries.*

At this point the notions involved in Theorem 2 of Chapter 1 have all been defined. In Theorems 3 and 4 we also use the notion of *weak Lie coordinatizability*, which involves a notion of Lie coordinatization in which the orientation condition is suppressed.

2.2 RANK

2.2.1 The rank function

Definition 2.2.1. *Let $D \subseteq \mathcal{M}$ be definable. A* rank function *(with finite values, or the symbol ∞ if undefined) is determined by the following conditions:*

1. *$rk\, D > 0$ if and only if D is infinite.*
2. *$rk\, D \geq n + 1$ if and only if there are definable D_1, D_2, π, f with $\pi : D_1 \to D$, $f : D_1 \to D_2$ such that*

   (i) *$rk\, \pi^{-1}(d) = 0$ for $d \in D$;*

   (ii) *$rk\, D_2 > 0$;*

   (iii) *$rk\, f^{-1}(d) \geq n$ for $d \in D_2$.*

If we are not in the $\aleph_0$-categorical case then these definitions should take place in a saturated model, and variations are possible using type-definable sets. We work in the $\aleph_0$-categorical setting. We write $rk(a/B)$ for the rank of the type of a over B, which is the minimum of $rk\, D$ for $a \in D$, D B-definable. In practice B is finite and the type reduces to the *locus* of a over B, which is the smallest B-definable set containing a.

Our definition of rank can be applied either to $\mathcal{M}$ or to $\mathcal{M}^{eq}$, and the latter is the more useful convention in the long run. When the distinction is significant, in connection with specific structures $\mathcal{M}$, we will refer to rk computed in $\mathcal{M}$ as *pre-rank*, and the rank computed in $\mathcal{M}^{eq}$ as *rank*.

Lemma 2.2.2

1. *$rk\, D = 0$ if and only if D is finite.*
1'. *$rk(a/B) = 0$ if and only if $a \in acl\, B$.*
2. *$rk(D_1 \cup D_2) = \max(rk\, D_1, rk\, D_2)$.*
2'. *(Extension property) If D is B-definable, then there is a complete type over B containing D and having the same rank.*
2''. *If $B_1 \subseteq B_2$ then $rk(a/B_2) \leq rk(a/B_1)$.*

Proof. Claims (1, 2) are straightforward and (1', 2') are direct consequences. Claim (2'') corresponds to the law: "if $D_1 \subseteq D_2$, then $rk\, D_1 \leq rk\, D_2$"; this is included in (2). ∎

Lemma 2.2.3. *Let $\mathcal{M}$ be $\aleph_0$-categorical. Then the following are equivalent for $a, b \in \mathcal{M}$:*

1. *$rk(a/b) \geq n + 1$.*
2. *There are a', c with $a' \in acl(abc) - acl(bc)$, and $rk(a/a'bc) \geq n$.*

Proof. Let D be the locus of a over b.

(1) $\implies$ (2). Suppose that $\pi : D_1 \to D$ has finite fibers, $f : D_1 \to D_2$ has fibers of rank at least n, and D_2 is infinite, with D_1, D_2, π, and f c-definable. Take $a' \in D_2 - acl(bc)$, and $a_1 \in f^{-1}(a')$ with $rk(a_1/a'bc) \geq n$ (using the Extension Property). Set $a_0 = \pi a_1$. Then we have $a' \in acl(a_0 bc) - acl(bc)$, and as $rk(a_1/a'bc) \geq n$ we find $rk(a_0/a'bc) \geq n$. Furthermore, as $tp(a_0/b) = tp(a/b)$ we can replace a_0 by a, replacing a', c by other elements.

(2) $\implies$ (1). Let a', c have the stated properties. Let D_1 be

$$\{(x, y) : tp(xy/bc) = tp(aa'/bc)\}$$

and let $\pi : D_1 \to D$, $f : D_1 \to D_2$ be the projections of D_1 onto the first and second coordinates, respectively. Then $f^{-1}(a')$ contains (a, a') and $rk(a/a'bc) \geq n$, so easily $f^{-1}(a')$ has rank at least n and hence the same applies to all fibers of f. It follows easily that D_1, D_2, f, π have the required properties for (1). $\blacksquare$

Lemma 2.2.4. *Let $\mathcal{M}$ be $\aleph_0$-categorical. If $rk(a/bc)$ and $rk(b/c)$ are finite, then $rk(ab/c)$ is finite and*

$$rk(ab/c) = rk(a/bc) + rk(b/c).$$

Proof. We use induction on $n = rk(a/bc) + rk(b/c)$, and the criterion of Lemma 2.2.3.

We show first that $rk(ab/c) \leq n$. Let d, e satisfy: $e \in acl(abcd) - acl(cd)$. We will show that $rk(ab/cde) < n$. We have either $e \in acl(abcd) - acl(bcd)$ or $e \in acl(bcd) - acl(cd)$ and correspondingly either $rk(a/bcde) < rk(a/bc)$ or $rk(b/cde) < rk(b/c)$. In either case induction applies to give $rk(ab/cde) < n$.

Now we show that $rk(ab/c) \geq n$. If $rk(b/c) = 0$ we observe that

$$rk(ab/c) \geq rk(a/c) \geq rk(a/bc) = n.$$

Assume $rk(b/c) > 0$, and take b', d with $b' \in acl(bcd) - acl(cd)$, such that $rk(b/b'cd) = rk(b/c) - 1$. Using the Extension Property we may suppose also that $rk(a/bb'cd) = rk(a/bc)$. By induction we find $rk(ab/b'cd) = n - 1$ and hence $rk(ab/c) \geq n$. $\blacksquare$

Corollary 2.2.5. *If $rk\, D = 1$, then acl defines a pregeometry on D, that is, a closure property of finite character with the exchange property.*

Definition 2.2.6. *We say that a and b are independent over C if*

$$rk(ab/C) = rk(a/C) + rk(b/C);$$

equivalently, $rk(a/bC) = rk(a/C)$.

Lemma 2.2.7. *The independence relation has the following properties:*

1. *Symmetry: If a and b are independent over C, then the same applies to b and a over C.*
2. *If a is algebraic over bC, then a is independent from b over C if and only if it is algebraic over C.*
3. *The following are equivalent:*

    (i) *a and bc are independent over E;*

    (ii) *a and b are independent over Ec, and a and c are independent over E.*

Proof. Each of these statements is clear on the basis of at least one of the criteria given in Definition 2.2.6.   ∎

This theory is relevant to our geometries, as they all have rank 1. This will be verified below.

2.2.2 Geometries

Lemma 2.2.8. *If J is a basic linear geometry then it has elimination of quantifiers.*

Proof. One checks that any suitably normalized atomic type is realized. In other words, using the basic universal axioms appropriate in each case, one shows that any existential formula in one variable can be reduced to a standard form, which is either visibly inconsistent or always realized. As we are dealing with basic geometries, the base field has been incorporated into the language, and we deal with structures whose underlying universe is of one of four types: degenerate, a vector space, a polar pair of spaces, or a quadratic pair (V, Q); these carry, variously, linear, bilinear, and quadratic structure. We may ignore the degenerate case and we defer the case of a quadratic geometry to the end. By taking the relevant bilinear or quadratic form to be identically zero in cases where it is not present, and expanding the domain of the type to a subspace B (or pair of subspaces in the polar case) which is nondegenerate whenever that notion is meaningful (this includes the polar case), we may assume the type to be realized has the following form:

(1) $x \notin B$
(2) $\beta(x, b) = \lambda(b)$
(3) $q(x) = \alpha$

The justification for (1) is that the excluded case is trivial, and the point of (3) is that any remaining conditions on q can be expressed in terms of the associated bilinear form in (2). Furthermore, the condition (2) is satisfied by an element of B, either because it is vacuous, or because B is nondegenerate, and after translation by such an element we get a similar system with $(1, 3)$ as above and (2) replaced by

$$(2') \qquad\qquad\qquad x \in B^{\perp}.$$

There is then nothing more to check unless q is nondegenerate. In this case one needs to know that q takes on all possible values in the orthogonal complement of any finite dimensional space. The following argument applies without looking at the classification of quadratic forms on finite dimensional spaces.

Let $K_0 \subseteq K$ be the set of values α such that q takes on the value α in the orthogonal complement to any finite dimensional space. Easily K_0 contains a nonzero element, is closed under multiplication by squares, and is closed under addition as $q(x + y) = q(x) + q(y)$ when x, y are orthogonal with respect to the associated bilinear form. It follows that $K_0 = K$.

Returning to the quadratic case, if the domain B of the type meets the set Q, then this is covered by the orthogonal case. Otherwise, we first add to the domain an element q of Q (we will have occasion later, in the treatment of imaginary elements, to revert to this point); the quantifier-free type of the extension is determined by the action of q on B, and the ω-invariant, both of which may be specified arbitrarily. ∎

Corollary 2.2.9. *The definable linear functions on the vector space V in a linear geometry are those afforded either by the inner product (if one is given, or is derivable from a quadratic form), or by the dual in the polar case.*

Proof. One checks that a definable subspace of finite codimension contains the kernel of a finite set of linear forms encoded directly in the structure (via a bilinear form, or polarity). Then any linear form whose kernel contains the kernels of these forms is expressible as a linear combination of them. ∎

Lemma 2.2.10. *The linear, affine, and projective geometries are all of prerank 1.*

Proof. It suffices to handle the basic linear case, and we can reduce the quadratic geometries to the orthogonal case. By quantifier elimination, algebraic closure is then linear span in the appropriate sense, which in the polar case takes place in two disjoint vector spaces. Thus the computation of rank is unaffected by the fact that the vector space structure may have been enriched. ∎

In the next section we discuss weak elimination of imaginaries, and one may then replace "pre-rank" by "rank" in the preceding.

Corollary 2.2.11. *If $\mathcal{M}$ is Lie coordinatizable, then $\mathcal{M}$ has finite rank, at most the height of the coordinatizing tree.*

Corollary 2.2.12. *If J is a linear, projective, or affine geometry, and a, b are finite sequences with $acl(a) \cap acl(b) = C$, then a and b are independent over C.*

Proof. Note that by definition the affine geometries include the linear model as a component. In the linear nonquadratic case we have noticed that the algebraic closure is the linear span; the analogous statement holds in the projective case. So in the linear and projective cases this is essentially a statement about linear algebra.

The affine and quadratic cases are similar: they may be expressed in the form (J_l, A), where J_l is linear and J_l (or in the polar case, part of J_l) acts regularly on the additional set A. A subspace is either an ordinary subspace of J_l (which may be polar) or a pair (B_l, B_a), where B_l is linear and B_a is an affine copy of B_l (with the usual modification in the polar case). If $acl(a) \cap acl(b)$ contains an affine (or quadratic) point then we are still essentially in the linear case; otherwise, we are working with affine dimension, which is 1 greater than the corresponding linear dimension. In this case it is important that $acl(a)$ and $acl(b)$ have a linear part determined by their affine parts (this should be rephrased slightly in the polar case, but the facts are the same). ■

2.2.3 A digression

The remainder of this section is devoted to additional remarks on rank notions which are far removed from our main topic.

Definition 2.2.13. *Let $\mathcal{M}$ be $\aleph_0$-categorical. Then ranks rk_α, taking values in $\mathbb{N} \cup \{\infty\}$, are defined as follows:*

1. $rk_0(D) = 0$ if D is finite, and is ∞ otherwise.
2. $rk_\alpha(D) > 0$ if and only if $rk_\beta(D) = \infty$ for $\beta < \alpha$.
3. $rk_\alpha(D) \geq n + 1$ if and only if there are $\pi : D_1 \to D$, $f : D_1 \to D_2$ definable, with

   (i) $rk_\alpha(\pi^{-1}(d)) = 0$ for $d \in D$;

   (ii) $rk_\alpha(f^{-1}(d)) \geq n$ for $d \in D_2$;

   (iii) $rk_\alpha(D_2) > 0$.

Remarks 2.2.14

1. In the superstable case, working in saturated models with type-definable sets, for D complete and α arbitrary there is a 0-definable quotient D' with $rk_\alpha D'$ finite and maximal. Writing $rk'_\alpha(D)$ for $rk_\alpha(D')$ we will have $U(D) = \sum_\alpha \omega^\alpha rk'_\alpha(D)$.

2. The ranks rk_α are additive and sets of α-rank 1 carry a geometry.

Definition 2.2.15. *If* $0 < rk_\alpha \mathcal{M} < \infty$, *we call* α *the* tier *of* $\mathcal{M}$. *According to the definition of* rk_α, *there is at most one tier for* $\mathcal{M}$.

Lemma 2.2.16. *There are pseudofinite* $\aleph_0$-*categorical structures of arbitrarily large countable tier, as well as structures of the same type with no tier.*

Proof. We deal first with countable tier. We have examples for $\alpha = 0$. In all other cases we proceed inductively, writing $\alpha = \sup(\beta_n + 1)$. We take countable pseudofinite $\aleph_0$-categorical structures D_n of tier β_n with $rk_{\beta_n}(D_n) \geq n(n+1)$ (replace D_n by a power if necessary) and encode them into D^{eq} for a new set D as follows.

We take initially a language L^* with sorts $D, D_1, D_2, \ldots$, whose restriction to D_n is the language of D_n. We also add generic maps $f_n : [D]^{n+1} \to D_n$; here the notation $[D]^i$ refers to unordered sets. The axioms are the axioms of D_n, relativized to that set, together with the following:

$$(*) \qquad \begin{array}{l} \text{For } t \in [D]^n \text{ and any } h_i : [t]^i \to D_i \text{ there is} \\ a \in D \text{ for which } f_i(s \cup \{a\}) = h_i(s) \text{ for } s \in [t]^i. \end{array}$$

This theory has D-quantifier elimination and is complete, consistent, and $\aleph_0$-categorical when interpreted as a theory of D, with D_n encoded in $\mathcal{M}^{\mathrm{eq}}$. For the finite model property, we begin with finite approximations to D_i for $i \leq N$, and we let D be large finite, f_n random; most choices satisfy $(*)$. As D^{n+1} maps onto D_n definably, we find $rk_{\beta_n} D \geq n$. Thus $rk_\alpha D \geq 1$; one can show $rk_\alpha D = 1$ and the tier is exactly α.

To get no tier we use sorts D_n and functions $f_n : D_n^n \to D_{n+1}^{n^2}$, satisfying the analog of $(*)$. Then $rk_\alpha(D_n) \geq n \, rk_\alpha D_{n+1}$ for all n and easily $rk_\alpha D_n = \infty$ for all n and α. We view this structure as encoded in D_1^{eq}. ∎

2.3 IMAGINARY ELEMENTS

Definition 2.3.1. $\mathcal{M}$ *has* weak elimination of imaginaries *if for all* $a \in \mathcal{M}^{\mathrm{eq}}$, *we have* $a \in dcl(acl(a) \cap \mathcal{M})$.

Lemma 2.3.2. *If* D *is* 0-*definable in* $\mathcal{M}$ *and* $D(a) = acl(a) \cap D$ *for* $a \in \mathcal{M}^{\mathrm{eq}}$, *then the following are equivalent:*

1. *D is stably embedded in $\mathcal{M}$ and admits weak elimination of imaginaries.*
2. *For $a \in \mathcal{M}^{\mathrm{eq}}$, $tp(a/D(a))$ implies $tp(a/D)$.*

Proof. $(1) \implies (2)$. Let $\varphi(x, y)$ be a formula with x a single variable (of the same sort as a). The relation $\varphi(a, y)$ defined on D is D-definable and hence has a canonical parameter d_0 in D^{eq}; note that $d_0 \in acl(a)$. By weak elimination of imaginaries there is $B \subseteq D(d_0) \subseteq D(a)$ such that $d_0 \in dcl \, B$ and hence $\varphi(a, y)$ is B-definable: $\varphi(a, y) \iff \varphi^*(b, y)$, with b in B. This last fact is part of $tp(a/B)$ and determines the φ-type of a over D. Thus (2) holds.

(2) $\implies$ (1). If $a \in D^{eq}$, then $a \in dcl(D)$, and hence by (2) we have $a \in dcl(D(a))$, as required for weak elimination of imaginaries.

Now suppose that $\varphi(x, a)$ is a formula implying $x \in D$, where x is a string of free variables. Let $A = D(a)$. If $tp(b/A) = tp(a/A)$, then $\varphi(x, a)$ and $\varphi(x, b)$ are equivalent, by (2). Thus $\varphi(x, a)$ is D-definable. ∎

Lemma 2.3.3. *Let J be a linear, projective, or affine geometry. Let $a \in J^{eq}$, and $A = acl(a) \cap J$. Then $acl(a) = acl(A)$.*

Proof. We may take J basic. Write $a = f(b)$ with b in J and f 0-definable. Take b' independent from b over $acl(a)$ in the sense of §2.2.1, and realizing the type $tp(b/ acl(a))$.

We claim that b and b' are independent over A. We have $a = f(b) = f(b')$ and thus $A \subseteq acl(b) \cap acl(b') \cap J \subseteq acl(a) \cap J = A$. Thus this reduces to Corollary 2.2.12.

Our two independence statements may be written out as follows:

$$rk(b'/Aab) = rk(b'/Aa); \quad rk(b'/Ab) = rk(b'/A).$$

Since $rk(b'/Aab) = rk(b'/Ab)$ and $rk(b'/Aa) = rk(b'a/A) - rk(a/A) = rk(b'/A) - rk(a/A)$, on comparing the two equations we find $rk(a/A) = 0$, and $a \in acl(A)$, as claimed. ∎

Corollary 2.3.4. *Let P be a projective geometry stably embedded in $\mathcal{M}$, A a subset of $\mathcal{M}$, and P_A the geometry obtained by taking acl relative to A as the closure operation. Then P_A is modular, i.e.,*

$$rk(ab) = rk(a) + rk(b) - rk(a \cap b)$$

for finite algebraically closed a, b.

Proof. By stable embedding and the preceding lemma we may replace A by $acl(A) \cap P$. ∎

Lemma 2.3.5. *Let J be a basic linear geometry. Then J has weak elimination of imaginaries.*

Proof. By the preceding lemma it suffices to prove the following: if $A \subseteq J$ is algebraically closed, $a \in J^{eq}$, and $a \in acl(A)$, then $a \in dcl(A)$.

We write $a = f(b)$ with f A-definable and $b = (b_1, \ldots, b_n)$, and we minimize n. Assuming $a \notin dcl(A)$, we have $n \geq 1$. Working over $A \cup \{b_1, \ldots, b_{n-1}\}$ we may suppose $n = 1$ and $b = b_n$. Let $D \subseteq J$ be the locus of b over A; of course, $b \notin A$. We examine the dependence of f on the element of D chosen.

Let $I = \{(x, y) \in D^2 : \langle xA \rangle \cap \langle yA \rangle = A\}$. The corner brackets are another notation for algebraic closure in J, intended to suggest linear span. For $(x, y) \in I$ the type of xy over A is determined by the inner product $\beta(x, y)$, with β nondegenerate or trivial, and possibly derived from a quadratic form; or else in the quadratic case, if $D \subseteq Q$, by $[x, y] = x(\sqrt{x + y})$. We will write $x \cdot y$ for the corresponding function in each case. So for some subset X of the field K we have

$$\text{For } (x, y) \in I, f(x) = f(y) \text{ if and only if } x \cdot y \in X.$$

Let $X_0 = K$ when we are dealing with a bilinear form, and $X_0 = \tau[K]$ with $\tau(x) = x^2 + x$ in the quadratic case with $D \subseteq Q$. Then in any case $X \subseteq X_0$ and it suffices to show that $X = X_0$, as then f is constant on independent pairs, and hence constant on D.

To see that $X = X_0$ it suffices to check that for $\alpha_{12}, \alpha_{13}, \alpha_{23} \in X_0$ there are x_1, x_2, x_3 independent over A for which $x_i \cdot x_j = \alpha_{ij}$ for $1 \le i < j \le 3$, as we then take $\alpha_{12} = \alpha_{23} \in X$ and $\alpha_{13} \in X_0$ arbitrary to conclude $X = X_0$. This is essentially a special case of the statement from which quantifier elimination was derived, though this was slightly obscured in the quadratic case by the suppression of some details.

We leave this calculation to the reader, but note that in the quadratic case, if the three elements x_1, x_2, x_3 are quadratic forms, we may write them as $q + \lambda_v^2, q, q + \lambda_w^2$, respectively, and find that the "target" values α_{ij} satisfy: $\alpha_{12} = q(v); \alpha_{23} = q(w);$ and

$$\alpha_{13} = (q + \lambda_v^2)(v + w) = \alpha_{12} + \alpha_{13} + \tau((v, w)). \qquad \blacksquare$$

Corollary 2.3.6. *Let J be a basic semiprojective geometry. Then J has weak elimination of imaginaries.*

Proof. Let $a \in J^{eq}$, let V be the vector space model covering J, and let $A = acl(a) \cap V$. Then $a \in dcl(A)$. Let $\mathbf{a}$ be a sequence of elements of J over which a is definable, and let $B = acl(a) \cap J$. The orbit of $\mathbf{a}$ in J over B is the same as its orbit over A, so $a \in dcl(B)$. $\qquad \blacksquare$

Remark 2.3.7. Projective geometries J need not have weak elimination of imaginaries, since the semiprojective geometry lies in J^{eq}.

Definition 2.3.8. *Let V be a vector space and A an affine V-space, with A and V definable in a structure $\mathcal{M}$. Let K be the base field.*
1. *A K-affine map $\lambda : A \to K$ is a map satisfying*

$$\lambda\left(\sum_i \alpha_i a_i\right) = \sum_i \alpha_i \lambda(a_i)$$

for scalars α_i with $\sum_i \alpha_i = 1$ (in which case the left side makes sense; linear operations make sense in A relative to a base point in A, and affine sums are independent of the basepoint).

2. *A^* is the set of $\mathcal{M}$-definable K-affine maps on A.*

Lemma 2.3.9. *In the notation of the previous definition, there is an exact sequence*

$$(0) \to K \to A^* \to V^* \to (0)$$

where V^* is the definable dual of V (consisting of all definable linear functionals).

Proof. K represents the set of constant functions. The map from A^* to V^* is defined as follows. For $\lambda \in A^*$ and $v \in V$, let $\lambda'(v) = \lambda(a+v) - \lambda(a)$, which is independent of the base point a. This is surjective since V^* lifts to A^* by choosing a base point in A.   ∎

Remarks 2.3.10

1. In this exact sequence it is possible that $A^* = K$ and $V^* = (0)$; indeed, this must occur in the stable case. V^* is described by the corollary to quantifier elimination in §2.2.1; in particular, V^* is definable.

2. Note that A^* is coded in $(V, V^*, A)^{\mathrm{eq}}$. The algebraic closure of an element of A^* in (V, V^*, A) will be the line in V^* generated by the corresponding linear map. For this reason we do not have weak elimination of imaginaries in (V, V^*, A). Note also that V^* is normally not mentioned explicitly, as it is identified with V when there is a nondegenerate bilinear map (assuming the situation is stably embedded).

3. We do have weak elimination of imaginaries in (V, V^*, A^*), as in the proof of Lemma 2.3.12 below, but this is not stably embedded in (V, V^*, A, A^*), as a base point in A gives a definable splitting of A^*—that is, a hyperplane complementary to the line of constants.

4. V^* is definable over A^*, so even in the polar case it is not necessary to include it in the geometry when A^* is present.

Lemma 2.3.11. *Let J be a basic, nonquadratic, linear geometry, and (J, A) a corresponding affine geometry. Then (J, A, A^*) admits quantifier elimination in its natural language.*

Proof. We take as the language the previous language for J, predicates for A and A^*, addition and subtraction maps $V \times A \to A$ and $A \times A \to V$, an evaluation map $A \times A^* \to K$, a K-vector space structure on A^*, distinguished elements of A^* corresponding to the constant functions, the canonical map $A^* \to V^*$ if V^* is present in some form, or an evaluation map $A^* \times V \to K$ if V^* is left to be encoded by A^*. As in the linear case we verify the realizability of suitably normalized atomic types. Since we can enlarge the domain of the types we can take a base point in A, identify A with V, and identify A^* with $K \oplus V^*$, putting us into the linear case. ∎

Lemma 2.3.12. *Let J be a basic nonquadratic linear geometry and let (J, A) be a corresponding basic affine geometry. Then (J, A, A^*) has weak elimination of imaginaries.*

Proof. As we have Lemma 2.3.3 for the affine case, and A^* is algebraic over V^*, we just have to check that the proof of Lemma 2.3.5 also goes through. As in that proof, our claim is that if $B \subseteq (J, A, A^*)$ is algebraically closed and $f : (J, A, A^*) \to (J, A, A^*)^{\text{eq}}$ is B-definable, then f is constant on each 1-type D over B.

We consider $I = \{(x, y) \in D^2 : \langle xB \rangle \cap \langle yB \rangle = B\}$, where the span is the algebraic closure in (J, A, A^*). (This includes the constant line in A^*.) We claim that f is constant along pairs in I; this will suffice. When $D \subseteq J$ it is convenient to view V^* as included in J, which is automatically the case except in the polar geometries. Then in dealing with D we may dispense with A and A^* and we are in the situation we treated previously. There remain the possibilities that $D \subseteq A$ or $D \subseteq A^*$.

Suppose that $D \subseteq A$. If B meets A, then we can replace D by a type realized in J. Suppose therefore that $B \cap A = \emptyset$. The type of D includes the values of affine maps on D and gives no further information about the type of a pair in I. Since the linear maps in B are covered by affine maps, this means that the only relevant part of B is $B \cap V$, and furthermore for $(x, y) \in D^2$, $x - y$ is orthogonal to $B \cap V$. Thus the type of such a pair, if it is not already determined, depends on the value of $Q(x - y)$ for a nondegenerate quadratic form Q. To repeat the previous argument we need independent elements v, w lying in $B^{\perp}$ with $Q(v), Q(w)$, and $Q(v + w)$ taking on arbitrary values. This we have.

Now suppose $D \subseteq A^*$. If B meets A, then A^* becomes identified with $K \oplus V^*$ and we return to the linear case. If $B \cap A = \emptyset$, then for $(x, y) \in I$ the type of the pair over B is determined by the type of the image in V^*, and we again return to the linear case. ∎

We now consider the relationship between the linear dual and the dual over the prime field. It turns out that the distinction is unimportant in the linear case but of some significance in the affine case.

Definition 2.3.13. *Let V be a vector space over the finite field F with prime field F_o, and A the corresponding affine space. We write $V^{*\circ}$ and $A^{*\circ}$ for the linear and affine dual with respect to the F_o-structure.*

Lemma 2.3.14. *There is a 0-definable group isomorphism τ between V^* and $V^{*\circ}$ given by $\tau f = \mathrm{Tr} \circ f$, and a 0-definable surjection $\tau^A : A^* \to A^{*\circ}$ given similarly by the trace $\mathrm{Tr} : F \to F_o$, with kernel the set of constant* *maps of trace 0.*

Proof. In the linear case, the two spaces have the same dimension over $F_\circ$. We check that the kernel of τ is trivial. Assume $\tau f = 0$. Then for any $v \in V$ and $\alpha \in F$, $\tau f(\alpha v) = Tr(\alpha f(v)) = 0$. As the trace form is nondegenerate on F, this means $f(v) = 0$.

In the affine case the difference in dimensions is the dimension of $F/F_\circ$ corresponding to the difference in the space of constant maps. As τ^A induces τ its kernel is contained in the space of constant maps.   ∎

We record the degree of elimination of imaginary elements afforded by $A^{*\circ}$.

Lemma 2.3.15. *Let (V, A) be a basic affine geometry, not of quadratic type. Let $C \subseteq (V, A, A^*)^{\mathrm{eq}}$ be definably closed and locally finite, that is, finite in each sort.*

If $acl(C) \cap (V \cup A) \subseteq C$, then $C = dcl(C \cap (V \cup A \cup A^{*\circ}))$.

Proof. Let $V_C = V \cap C$, $A_C = A \cap C$, $A^*_C = A^* \cap acl(C)$. By weak elimination of imaginaries $C = dcl(acl(C) \cap (V \cup A \cup A^*)) = dcl(V_C \cup A_C \cup A^*_C)$. As V^* is identified with a quotient of $A^{*\circ}$ it will suffice to check that

$$Mult(A^*_C/C) = Mult(A^*_C/C \cap A^{*\circ}, C \cap V^*).$$

Let $v^*_1, \ldots, v^*_d$ be a basis for $C \cap V^*$ and let a^*_i be a lifting of v^*_i to A^*. The element a^*_i is chosen from an affine line over the base field F. We have for each i

$$Mult(a^*_i/a^*_1, \ldots, a^*_{i-1}, C) \leq Mult(a^*_i/a^*_1, \ldots, a^*_{i-1}, V^* \cap C, A^{*\circ} \cap C)$$

and it suffices to show equality.

Let $K = Aut(a^*_i + F/a^*_1, \ldots, a^*_{i-1}, C)$, a subgroup of $(F, +)$. Let L be the space of K-invariant affine maps over $F_\circ$ on $a^*_i + F$. We have $Aut(a^*_i + F/L) = K$, since a translation $x \rightarrow x + \alpha$ on $a^*_i + F$ leaves L invariant if and only if the linear maps induced by L annihilate α, and these are just the $F_\circ$-linear maps annihilating K. Accordingly, for $A^*_i = \langle a^*_1, \ldots, a^*_i \rangle$ we have

$$Aut(A^*_i/a^*_1, \ldots, a^*_{i-1}, C) = Aut(A^*_i/v_i, L).$$

Now $L \subseteq dcl(a^*_1, \ldots, a^*_{i-1}, C) \cap (a^*_1 + F)^{*\circ}$, and we need

$$L \subseteq dcl(a^*_1, \ldots, a^*_{i-1}, C) \cap A^{*\circ}.$$

For $f \in L$ inducing the linear map $\bar{f}$ and $a^* \in a^*_i + F$ define $f_{a^*} \in A^{*\circ}$ by $f_{a^*}(a) = \bar{f}((a, a^*)) - f(a^*)$. This does not depend on the choice of a^*: $f_{a^*+\alpha}(a) = \bar{f}((a, a^*) + \alpha) - [f(a^*) + \bar{f}(\alpha)] = f_{a^*}(a)$. Thus f defines $f' = f_{a^*}$ and the converse is obvious.   ∎

Lemma 2.3.16. *Let V be part of a stably embedded basic linear geometry J with base field F. Let A be an affine space over V. Assume A and V are 0-definable. Then there is a 0-definable space, which we will denote FA, such that FA contains V as a subspace of codimension 1, and A as a coset of V. The space FA is unique up to canonical definable isomorphism.*

Proof. We let FA be $F \times A \times V$ modulo the equivalence relation defined by: $(\alpha, a, v) \sim (\alpha', a', v')$ if and only if $\alpha = \alpha'$, $\alpha(a - a') = v - v'$. Equivalence classes will be denoted in terms of their representatives as $\alpha a + v$. The scalar multiplication will be defined by

$$\beta(\alpha a + v) = (\beta \alpha)a + \beta v.$$

This is clearly well defined.

To define addition on FA, note that for any $a_\circ \in A$ the elements of FA are uniquely representable in the form $\alpha a_\circ + v$. Thus we may set

$$(\alpha a_\circ + v) + (\alpha' a_\circ + v') = (\alpha + \alpha')a_\circ + (v + v').$$

This definition is immediately seen to be independent of the choice of $a_\circ$. Thus the construction is 0-definable. One checks the vector space axioms. Evidently V sits as a subspace of codimension 1 and A as a coset.

Verification of the uniqueness statement is straightforward. ∎

Lemma 2.3.17. *Let V be a nonquadratic basic linear geometry, possibly with distinguished elements, forming part of a geometry J with field of scalars F which is stably embedded in $\mathcal{M}$. Let A be a C-definable affine space over V. Then*

1. *$FA \cup J$ and $FA \cup (FA)^* \cup J$ are stably embedded.*
2. *Suppose A is not in $acl(J, C)$ and let $C_\circ$ be*

$$[acl(C) \cap J] \cup [acl(C) \cap A] \cup [acl(C) \cap A^{*\circ}].$$

Then $(J, FA, FA^{*\circ}, C_\circ)$ with its intrinsic geometric structure is fully embedded in $\mathcal{M}$ over $C \cup C_\circ$.

Proof.

Ad 1. For $a \in A$ we have $FA \subseteq dcl(a, V)$. Furthermore, $(FA)^* \subseteq dcl(a, V^*)$ since $f \in (FA)^*$ is determined by its restriction to V and its value at a. Thus this is immediate.

Ad 2. Let $\mathcal{N}$ be $(J, FA, FA^{*\circ}, C_\circ)$ with its intrinsic geometric structure, and let $\mathcal{N}'$ be $\mathcal{N}$ with its full induced structure. As $\mathcal{N}$ is stably embedded, any 0-definable set D in $\mathcal{N}'$ is definable in $\mathcal{N}$ with parameters. We claim that D is 0-definable in $\mathcal{N}$.

Let d be the canonical parameter for D in $\mathcal{N}$, and $d' = [acl(d) \cap (A \cup J)] \cup [dcl(d) \cap A^{*\circ}]$. By Lemma 2.3.15 $d \in dcl(d')$ in $\mathcal{N}$. In $\mathcal{N}'$ by assumption $d' \in dcl(\emptyset)$, and thus $d' \in C_\circ$. Thus D is 0-definable in $\mathcal{N}$. ∎

Lemma 2.3.18. *Let V be a nonquadratic basic linear geometry, possibly with distinguished elements, forming part of a geometry J with field of scalars F which is stably embedded in $\mathcal{M}$. Let A be a C-definable affine space over V. Let $C' \supseteq C$ with $acl(C') \cap (J \cup A^{*\circ}) \subseteq C'$ and $acl(C', J) \cap A = \emptyset$. Then for $a \in A$, $tp(a/C' \cap A^{*\circ})$ implies $tp(a/C')$.*

Proof. We may take $C = \emptyset$. By the preceding lemma $A \cup A^* \cup J$ is fully embedded in $\mathcal{M}$ over the parameters $C_\circ = C' \cap (J \cup A^{*\circ})$. Thus

$$tp(a/\, dcl(C') \cap (A, A^*, J)^{\mathrm{eq}}) \implies tp(a/C').$$

By Lemma 2.3.15

$$dcl(C') \cap (A, A^*, J)^{\mathrm{eq}} \subseteq dcl(C_\circ).$$

By quantifier elimination, $tp(v/C' \cap A^{*\circ})$ determines $tp(v/C' \cap A^{*\circ}, a)$ for $v \in J$, so $tp(a/C' \cap A^{*\circ})$ determines $tp(a/C' \cap A^{*\circ}, J)$. The claim follows. ∎

Lemma 2.3.19. *A Lie coordinatizable structure is $\aleph_0$-categorical.*

Proof. It suffices to treat the case of a structure $\mathcal{M}$ equipped with a Lie coordinatization. The argument is inductive, using Lemmas 2.3.5 and 2.3.12 with Lemma 2.3.2, and some control of the algebraic closure. Let $\mathcal{N}_h$ be the part of $\mathcal{M}$ coordinatized by the tree up to height h, let $\mathcal{N}$ be $\mathcal{N}_h$ together with finitely many coordinatizing geometries at height $h + 1$, and let J be a further coordinatizing geometry at any level, with defining parameter a. Our claim is:

J realizes finitely many types over any finite subset of $\mathcal{N} \cup \{a\}$.

In the main case, J is itself at height $h+1$ and thus a is already in $\mathcal{N}_h$. However, with J fixed, we proceed inductively on h and on the number of components at level $h + 1$, beginning with $\mathcal{N}$ empty.

Given this result, one can then get the uniform bounds required for $\aleph_0$-categoricity by one more induction over the tree (by height alone).

It will be convenient to assume that the geometries involved are basic, and are either finite, linear, or affine; that is, projective geometries should be replaced by their linear covers. This cannot be done definably. Since the expanded version of $\mathcal{M}$ interprets $\mathcal{M}$ and has essentially the same coordinatizing tree, this implies the stated result for $\mathcal{M}$.

Since the case in which J is finite is trivial, we need deal only with the linear and affine cases, to which Lemmas 2.3.5 and 2.3.12 apply, and may be combined with Lemma 2.3.2. This reduces the problem to the following: for $A \subseteq \mathcal{N}$ finite, show that $acl(Aa) \cap J$ is finite.

Suppose on the contrary that $acl(Aa) \cap J$ contains arbitrarily large finite-dimensional Aa-definable subspaces V of J. Fix such an Aa-definable subspace V of J. $\mathcal{N}$ is B-definable for some set of parameters B lying in $\mathcal{N}_h$, and by induction $acl(Ba) \cap J$ is finite. As A' varies over the set of realizations in $\mathcal{N}$ of the type of A over Ba, the corresponding A'-definable subspace V' varies over the realizations of the type of V over $acl(Ba) \cap J$. Let n_1 be the number of types of sets AA' as A' varies in this manner, and let n_2 be the number of types of the corresponding sets VV'. Now n_1 is bounded, by induction hypothesis, since $B \subseteq \mathcal{N}_h$ and $A \subseteq \mathcal{N}$; $\mathcal{N}$ can be thought of as obtained by appending one geometry J' to a structure $\mathcal{N}'$ with $\mathcal{N}_h' = \mathcal{N}_h$ and with one fewer component at height $h + 1$. We have arrived at the following: n_1 is bounded, and as the dimension of V increases, n_2 is unbounded; but $n_1 \geq n_2$. This contradiction yields a bound on the dimension of V and hence on the size of $acl(Aa) \cap J$. ∎

2.4 ORTHOGONALITY

Definition 2.4.1

1. *A normal geometry is a structure J with the following properties (uniformly—in every elementary extension):*

   (i) *$acl(a) = a$ for $a \in J$.*
   (ii) *Exchange: if $a \in acl(Ba') - acl(B)$, then $a' \in acl(Ba)$.*
   (iii) *If $a \in J^{\mathrm{eq}}$, then $acl(a) = acl(B)$ for some $B \subseteq J$.*
   (iv) *For $J_0 \subseteq J$ 0-definable and nonempty, if $a, a' \in J$ and $tp(a/J_0) = tp(a'/J_0)$ then $a = a'$.*

2. *A normal geometry is* reduced *if it satisfies the further condition:*

   (v) *$acl(\emptyset) = dcl(\emptyset)$ in J^{eq}.*

   This distinction is illustrated by Example 2.4.5.

Lemma 2.4.2. *Projective geometries in our sense are normal geometries. The basic projective geometries are normal and reduced.*

Proof. Note that we include the polar and quadratic cases.

Conditions (i) and (ii) are the usual geometric properties in most cases. In the polar and quadratic case this includes the fact that the various parts of the geometry do not interact pointwise, e.g., for $q \in Q$ in the quadratic case, $acl(q) \cap V = \emptyset$. This can be computed in the basic linear model using quantifier elimination. We remark also that (i) requires $acl(\emptyset) = \emptyset$, which is not so much true as a matter of how the structure is viewed; for this purpose one takes a model in which objects such as the field K are encoded in $\mathcal{M}^{\mathrm{eq}}$ (or in the language). Condition (iii) was verified in §2.3. For (iv), note that apart from the polar and quadratic cases, if there are nontrivial 0-definable subsets they are determined by the set of values of a quadratic form or a hermitian form on the line representing a projective point. If a and a' have the same type over J_0, then they lift to points in the linear space having the same type over the preimage of J_0. But these sets generate the whole vector space.

In the polar case it may happen (e.g., in the basic case) that the two vector spaces involved are 0-definable. However, the type of a linear form over a vector space determines the linear form. Similarly in the quadratic case, the type of a quadratic form over its domain determines the form, and conversely the type of a vector over Q determines the vector as an element of the dual space, and hence determines the vector.

For (v) in the basic case, apply weak elimination of imaginaries for the associated basic semiprojective geometry to an element of $acl(\emptyset)$.   ∎

The following is a modified form of Lemma 1 of [HrBa].

Lemma 2.4.3. *Let J_1, J_2 be normal geometries which are fully embedded and 0-definable in a structure $\mathcal{M}$. Then one of the following occurs:*

1. *J_1, J_2 are* orthogonal*: every 0-definable relation on $J_1 \cup J_2$ is a boolean combination of sets of the form $R_1 \times R_2$ with R_i an $acl(\emptyset)$-definable relation on J_i; or*
2. *J_1, J_2 are* 0-linked*: there is a 0-definable bijection between J_1 and J_2.*

Proof. If (1) fails then there is a 0-definable relation $R \subseteq J_1^{n_1} \times J_2^{n_2}$ for some n_1, n_2 which is not a finite union of $acl(\emptyset)$-definable rectangles $A_1 \times A_2$ ($A_i \subseteq J_i^{n_i}$ $acl(\emptyset)$-definable). It follows by compactness that we have $b_1 \in J_1^{n_1}$ such that $R(b_1) = \{b_2 \in J_2^{n_2} : R(b_1, b_2)\}$ is not $acl(\emptyset)$-definable. Our first claim is

(∗)   If $b_1 \in J_1^{n_1}$, $R \subseteq J_1^{n_1} \times J_2^{n_2}$ is 0-definable, and $R(b_1)$ is not $acl(\emptyset)$-definable, then $acl(b_1)$ meets J_2.

By stable embedding, $R(b_1)$ is J_2-definable. Let $c_2 \in J_2^{eq}$ be its canonical parameter. Then by assumption c_2 is not algebraic over $\emptyset$, and then by (iii) we conclude that $acl(c_2)$ meets J_2, and (∗) follows.

Now take $a_2 \in acl(b_1) \cap J_2$ and let $S(a_2)$ be the locus of b_1 over a_2. As a_2 is algebraic over $S(a_2)$, $S(a_2)$ is not definable over $acl(\emptyset)$. Thus by another application of (∗), $acl(a_2)$ meets J_1. Let $a_1 \in acl(a_2) \cap J_1$. By the argument just given, we can also find $a_2' \in acl(a_1) \cap J_2$. But then $a_2' \in acl(a_2) \cap J_2 = \{a_2\}$ and thus $acl(a_1) = acl(a_2)$, and furthermore we have shown that in this relation a_1 determines a_2 (and of course, conversely). Thus $dcl(a_1) = dcl(a_2)$ and we have a 0-definable bijection f between two 0-definable sets $D_1 \subseteq J_1$ and $D_2 \subseteq J_2$. By (iv) and compactness each element a_1 of J_1 is determined by some a_1-definable subset of D_1, and hence (using f) by some a_1-definable subset $T(a_1) \subseteq D_2$. Therefore this set $T(a_1)$ is not definable over $acl(\emptyset)$, and by (∗) and the subsequent argument a_1 belongs to the domain of some

0-definable bijection between parts of J_1 and J_2. By compactness J_1 and J_2 are 0-linked. ∎

Remark 2.4.4. Under the hypotheses of the preceding lemma, if J_1 and J_2 are reduced, then the first alternative can be strengthened as follows:

1'. J_1, J_2 are *strictly orthogonal*: every 0-definable relation on $J_1 \cup J_2$ is a boolean combination of sets of the form $R_1 \times R_2$ with R_i a 0-definable relation on J_i.

This holds since $dcl(\emptyset)$-definable sets are 0-definable.

Example 2.4.5. J_1, J_2 *carry equivalence relations* E_1, E_2 *with two infinite classes and no other structure. Then these are normal geometries, but not reduced. In* $J_1 \times J_2$ *we may add a bijection between* J_1/E_1 *and* J_2/E_2. *This would fall under the orthogonal case, but not the strictly orthogonal case.*

Lemma 2.4.6. *Let* J_1 *and* J_2 *be basic linear geometries canonically embedded in the structure* $\mathcal{M}$. *Suppose that in* $\mathcal{M}$ *there is a 0-definable bijection* $f : PJ_1 \leftrightarrow PJ_2$ *between their projectivizations. Then there is a 0-definable bijection* $\hat{f} : J_1 \leftrightarrow J_2$ *which is an isomorphism of unoriented weak geometries, and which induces* f.

Proof. Without loss of generality we may take the universe to be $J_1 \cup J_2$. Recall that in the basic linear geometries any bilinear or quadratic forms involved may be taken to be K-valued, and $acl(\emptyset) = dcl(\emptyset)$.

J_i consists either of a single vector space, a pair of spaces in duality, or a quadratic geometry (V, Q) and correspondingly PJ_i consists either of the corresponding projective model, two projective spaces, or the pair (PV, Q). The given f preserves algebraic closure, which is the span in the projective sense (except in Q) and hence f is covered by a map $\hat{f}$ which is linear on each vector space in J_i (relative to an isomorphism of the base fields) and which agrees with f on Q in the quadratic case. At this point we will identify the base fields, writing $K = K_1 = K_2$. There are finitely many such maps $\hat{f}$, and the set of them is implicitly definable, so by Beth's theorem they are definable over $acl(\emptyset) = dcl(\emptyset)$, or in other words, are 0-definable.

Fix one such $\hat{f}$. The type of $\hat{f}(a)$ is determined by the type of a, for a a finite string of elements. When a quadratic form is present we may recognize the totally isotropic spaces as those on which only one nontrivial 1-type is realized; in the polar case a totally isotropic space consists of a pair of orthogonal spaces, and one nontrivial 1-type is realized in each factor. It follows that $\hat{f}$ preserves orthogonality. Furthermore, if quadratic or skew quadratic forms Q_1, Q_2 are present (given, or derived from a hermitian form), then there is a function F for which $Q_2(\hat{f}(x)) = F(Q_1(x))$, where $F : K_0 \to K_0$ with $K_0 = K$ except in the hermitian case, where it is the fixed field of an automorphism σ of order 2.

The function F is additive (consider orthogonal pairs) and linear with respect to elements of K^2 or in the hermitian case, K_0. In any case, it follows that F is linear on K_0 and is given by multiplication by an element α of K_0; in other words, $Q_2 = \alpha Q_1$. This sort of shift is allowed by a weak isomorphism, so our claim follows except in the polar, symplectic, and quadratic cases, to which we now turn.

In the polar and symplectic cases the 1-type structure is trivial and we have a function $F : K \to K$ for which $\beta_2(\hat{f}v, \hat{f}w) = F(\beta_1(v, w))$, where β_i gives either a duality between two spaces, or a symplectic structure. This map is visibly linear, so we are in the situation considered previously.

Now we consider the quadratic case. On the symplectic part we have $\beta_{V_2} = \alpha\beta_{V_1}$ for some α. Considering pairs (v, q) in $V \times Q$ we find that $\hat{f}q(\hat{f}v) = F'(q(v))$ for some function F' which similarly must be multiplication by a constant (for example, by considering the effect of replacing v by a scalar multiple); as the form associated to $\hat{f}q$ is $\alpha \cdot \beta_{V_1}$, we find $\hat{f}q = \alpha q$. This leads to the particularly unsatisfactory conclusion that the actions of V_1 and V_2 on Q_1 and Q_2 are related by

$$\hat{f}q +_2 \hat{f}v^2 = \hat{f}(q +_1 (\alpha^{1/2}v)^2).$$

We can, however, adjust $\hat{f}$ by taking $\hat{f}'(v) = \alpha^{1/2}v$ and then we find that the inner product, the action of Q, and translation by V all agree in the two models. ∎

Lemma 2.4.7. *Let J_1 and J_2 be basic quadratic or polar linear geometries canonically embedded in the pseudofinite structure $\mathcal{M}$. Suppose that in $\mathcal{M}$ there is a 0-definable bijection $f : PJ_1 \leftrightarrow PJ_2$ between the projectivizations of V_1 and V_2 (V_i is one of the two factors of J_i, in the polar case, and the vector part, in the quadratic case). Then there is a 0-definable bijection $\hat{f} : J_1 \leftrightarrow J_2$ which is an isomorphism of weak geometries, and which induces f.*

Proof. By the preceding lemma, f lifts to the linear part of J_1, J_2 covering PV_i. It remains to be seen that the linear or quadratic forms on V_2 which correspond to elements of J_1, transported by $\hat{f}$, are realized by elements of J_2. In finite approximations to $\mathcal{M}$, *all* such maps are realized, and in particular, all definable ones are realized in $\mathcal{M}$ by elements of J_2. If $\hat{f}$ is chosen to preserve the symplectic structure (exactly) in the quadratic case, then all structure will be preserved by the induced map. ∎

Lemma 2.4.8. *Let $\mathcal{M}$ be a structure, D and I definable subsets of $\mathcal{M}$, and $\{A_i : i \in I\}$ a collection of uniformly i-definable subsets of $\mathcal{M}$. Assume that $\mathrm{acl}(\emptyset) = \mathrm{dcl}(\emptyset)$ in D^{eq}, and that D is orthogonal to I and is orthogonal to each A_i over i. Then D is orthogonal to $I \cup \bigcup_i A_i$ (and hence strictly orthogonal to the same set).*

Proof. It will be convenient to use the term "relation between A and B" for a subset of $A^m \times B^n$ with m, n arbitrary. We have to analyze a relation between D and $I \cup \bigcup_i A_i$, and it suffices to consider the part relating D to $\bigcup_i A_i$. Fix i. Then R gives a relation between D and A_i involving a finite number of $acl(i)$-definable subsets of D^m for some m. These belong to a finite i-definable boolean algebra of subsets of D^m, which by strict orthogonality is 0-definable over D, and may be taken to be independent of i by dividing I into 0-definable sets. The elements of this boolean algebra belong to $acl(\emptyset)$ in D^{eq} and hence are 0-definable. The relation with A_i can be expressed in terms of them, and I may be broken up further into finitely many 0-definable pieces on which the definition is constant.   ∎

Definition 2.4.9. *The* localization P/A *of a projective geometry P over a finite set A is the geometry obtained from the associated linear geometry L as follows. Let $L_0 = acl(A) \cap L$, and projectivize $L_0^\perp / rad(L_0)$. If the vector space $L_0^\perp / rad(L_0)$ supports a quadratic geometry Q_0 then add that geometry to the localization as well. (The radical $rad(L_0)$ is $L_0 \cap L_0^\perp$; in the quadratic case $L_0^\perp$ has a quadratic part which is taken to consist of quadratic forms which vanish on $rad(L_0)$; in the orthogonal case in characteristic 2 we may also have to add a quadratic part—see the following remark.)*

Remark 2.4.10. The previous definition uses the convention that inner products are 0 where undefined. In the linear case one therefore works with L/L_0. In the polar case L_0 consists of two spaces and the orthogonal spaces "switch sides." In the quadratic case $Q/Q \cap L_0^\perp$ is a space of quadratic forms on the correct space $(L \cap L_0^\perp)/ rad(L_0)$. (It would not be well-defined, however, as a space of forms on $L/ rad(L_0)$.) Finally, one unusual phenomenon occurs in localizing orthogonal geometries in characteristic 2. Let q be a quadratic form associated to a nondegenerate symplectic form on V, and for simplicity let $A = \{v\}$ be a single nonzero vector of V. If $q(v) = 0$, then the form q descends to $\bar{L} = v^\perp / \langle v \rangle$; otherwise, for each nonsingular 2-space H containing $\langle v \rangle$ in L, the restriction of q to $H^\perp$ induces a quadratic form on $\bar{L}$, and as H varies the collection $\bar{Q}$ enlarges $\bar{L}$ to a quadratic geometry $(\bar{L}, \bar{Q})$.

If P is a basic projective geometry, then this geometry is again a basic projective geometry, since the base field is named. In most cases it gives a geometry of the same type we began with. We could also define the *full localization* by taking P modulo the equivalence relation $acl(xA) = acl(yA)$, with all induced structure. The nontrivial atoms of the full localization are either components of our localization or affine spaces over its linear part.

Lemma 2.4.11. *Let P, Q be basic projective geometries defined and orthogonal over the set A and fully embedded over A in $\mathcal{M}$. Then their localizations are orthogonal over any set B over which they are defined.*

Proof. We may suppose that $A \subseteq B$ or $B \subseteq A$, with the proviso in the latter case that we allow P, Q to be localizations of geometries defined over B.

If $A \subseteq B$ and $P/B, Q/B$ are nonorthogonal, then they have a B-definable bijection which is unique and hence defined over $A \cup (acl(B) \cap (P \cup Q))$ (which serves to define the localizations). But over A this gives a relation on $P \cup Q$ which violates the orthogonality.

If $B \subseteq A$ and $P = \hat{P}/A$, $Q = \hat{Q}/A$ with $\hat{P}, \hat{Q}$ basic B-definable projective geometries, then nonorthogonality over B gives a B-definable bijection between $\hat{P}$ and $\hat{Q}$ which induces an A-definable bijection between the localizations. ∎

2.5 CANONICAL PROJECTIVE GEOMETRIES

Throughout this section we work in a Lie coordinatized structure $\mathcal{M}$.

Definition 2.5.1. *Let $J = J_b$ be a b-definable weak projective Lie geometry in the structure $\mathcal{M}$. Then J is a* canonical projective *geometry if*

1. *J is fully embedded over b; and*
2. *If $tp(b') = tp(b)$ and $b' \neq b$, then J_b and $J_{b'}$ are orthogonal.*

A terminological note: there is no connection between the use of the term "canonical" in connection with canonical embeddings, and canonical projectives. In the case of embeddings the term refers to the so-called "canonical language," which has not been introduced explicitly here, while in the latter case it refers to the canonicity condition (2).

Lemma 2.5.2. *Let P_b be a b-definable projective geometry fully embedded in a Lie coordinatizable structure $\mathcal{M}$. Then there is a canonical projective geometry in $\mathcal{M}^{eq}$ nonorthogonal to P_b over a finite set.*

Proof. We may assume P_b is basic, and since it lives in $\mathcal{M}^{eq}$, we may replace $\mathcal{M}$ by a bi-interpretable structure and suppose that $\mathcal{M}$ is coordinatized by Lie geometries. If P_b is orthogonal to each of the coordinatizing geometries over their defining parameters, then repeated use of Lemma 2.4.8 shows that P_b is orthogonal to the ambient model $\mathcal{M}$, and hence to itself, which is not the case (the equality relation refutes this).

So we may suppose that P_b is one of the coordinatizing geometries, and that b is the parameter associated with P_b in the coordinatization of $\mathcal{M}$, so that it represents a branch $(b_1, \ldots, b_n)$ (or $b_0, \ldots, b_n$ with b_0 the 0-definable root) of the tree structure on $\mathcal{M}$ associated with a sequence of geometries (and finite algebraically closed sets) in $\mathcal{M}$, with $b_n = b$. Minimize n subject to nonorthogonality to the original geometry, so that for each geometry of the form J_{b_i}, with $i < n$, the associated projective geometry is orthogonal to P_b.

Consider the conjugates $P_{b'}$ of P_b. If $P_b, P_{b'}$ are nonorthogonal over a finite set, then the appropriate localization of $P_{b'}$ is orthogonal to the coordinatizing geometries Q for b over any set over which $P_{b'}$ and Q are defined. It follows by induction that $P_{b'} \cap acl(b', b_1, \ldots, b_i) = \emptyset$ for all $i \le n$; notice that the induction step is vacuous when b_i is algebraic over its predecessor. For $i = n$ we have $acl(b, b') \cap P_{b'} = \emptyset$ and similarly $acl(b, b') \cap P_b = \emptyset$. Thus the nonorthogonality gives a unique (b, b')-definable bijection between P_b and $P_{b'}$, preserving the unoriented weak structure, and also, by an explicit hypothesis, preserving the Witt defect in the quadratic case.

Nonorthogonality of the associated geometries defines an equivalence relation on the conjugates of b and for any pair b', b'' of equivalent conjugates we have a canonical (b', b'')-definable isomorphism $\iota_{b', b''}$ between the geometries as weak geometries. Let b_1, b_2, b_3 be three conjugates of b for which the corresponding geometries are nonorthogonal. Using the orthogonality as above we may show that $acl(b_1, b_2, b_3) \cap P_{b_3} = \emptyset$ and hence the unique $acl(b_1, b_2, b_3)$-definable bijection between P_{b_1} and P_{b_3} agrees with the composition of the canonical bijections $P_{b_1} \leftrightarrow P_{b_2}$ and $P_{b_2} \leftrightarrow P_{b_3}$. So these identifications cohere and we can attach to an equivalence class c of conjugates of b a single weak projective geometry Q_c^w canonically identified with the given weak projective geometries. The geometry we want is the basic projective geometry Q_c associated with Q_c^w. We still must check that it is in fact canonical. This follows since the conjugates of c distinct from c are the classes of conjugates of b inequivalent to b. ∎

Lemma 2.5.3. *Let P_b be a b-definable projective geometry fully embedded in a Lie coordinatizable structure, and let J_c be a canonical projective geometry nonorthogonal to P_b with canonical parameter c. Then $c \in dcl(b)$ and $P_b \subseteq dcl(b, J_c)$.*

Proof. For the first point, if c' is a conjugate of c over b, then $P_{c'}$ is nonorthogonal to P_b and hence to P_c; so $c = c'$. Thus $c \in dcl(b)$. There is a (b, c)-definable bijection between the localizations of P_b and J_c, and the localization of P_b over $\{b, c\}$ is P_b since $c \in dcl(b)$ (or for that matter since $c \in acl(b)$). Thus this bijection induces a function from J_c onto P_b. ∎

Lemma 2.5.4. *Let P_b, $P_{b'}$ be b-definable and b'-definable canonical projective geometries, respectively, not assumed to be conjugate. If P_b and $P_{b'}$ are nonorthogonal, then $dcl(b) = dcl(b')$ and there is a unique (b, b')-definable bijection between them, which is an isomorphism of weak, unoriented geometries.*

Proof. The first point follows from Lemma 2.5.3 and allows us to construe (b, b') as either b or b'. The rest is in Lemmas 2.4.3 and 2.4.6. ∎

We will discuss the issue of orientation further.

Lemma 2.5.5. *Let P_b be a canonical projective quadratic geometry. There is a coordinatizing quadratic geometry J_c and a definable unoriented weak isomorphism of P_b with J_c. We may choose c so that if we orient P_c according to this isomorphism, the orientation is independent of the choice of c within its type over b.*

Proof. Let J_c be a coordinatizing geometry not orthogonal to P and minimized in the sense that c is as low in the tree structure on $\mathcal{M}$ as possible. Then by the previous lemma $b \in dcl(c)$ and by the minimization, as in the proof of Lemma 2.5.2, $acl(c) \cap P = \emptyset$. Thus the nonorthogonality gives a definable weak unoriented isomorphism. Conjugates of c over b, or for that matter conjugates of c over the empty set for which the corresponding geometry is nonorthogonal to P_b, have compatible orientations by the orientation condition in the definition of Lie coordinatization. ∎

For a discussion of orientation the following terminology is convenient.

Definition 2.5.6
1. *A standard system of geometries is a 0-definable function $J : A \to \mathcal{M}^{eq}$ whose domain A is a complete type over $\emptyset$ and whose range is a family of canonical projective geometries.*
2. *Two standard systems of geometries are equivalent if they contain a pair of nonorthogonal geometries. In this case there is a 0-definable identification between the systems, since nonorthogonality gives us a 1-1 correspondence between the domains, and the nonorthogonal pairs have canonical identifications.*

Lemma 2.5.7. *In a Lie coordinatized structure the quadratic geometries can be assigned compatible orientations, in the sense that in nonorthogonal geometries the orientations are identified by the canonical weak unoriented isomorphism between appropriate localizations. This can be done 0-definably.*

Proof. We first orient the standard systems made up of projective quadratic geometries. Here we just choose one representative of each equivalence class of such systems, and use the given orientations.

With this as a frame of reference we orient an arbitrary quadratic geometry P_b. There is a unique canonical projective quadratic geometry J_c oriented in the first step and nonorthogonal to P_b, and we have $c \in dcl(b)$. There is a canonical isomorphism between P_b and the localization of J_c at $A = acl(b) \cap J_c$. (By Lemma 2.5.5 it provides a well-defined identification of orientations.) It will be convenient to look at the linear quadratic geometry (V, Q) associated with J_c, and at $B = acl(b) \cap (V, Q)$, which carries the same information as A (as far as J_c is concerned).

B does not meet Q, as this would result in the localization of J_c at B, and hence also P_b, being orthogonal rather than quadratic. Let B_0 be a linear complement to $rad\,B$ in B. We can localize at B in two steps: first with respect to B_0, then with respect to $rad\,B$. At the first step the set Q is unchanged, but we modify the Witt defect as follows: $\omega^{B_0}(q^-B_0^\perp) = \omega(q) + \omega(q^-B_0)$. Here, on the right, ω is in one case the orientation function chosen already on J_c, and in the other the ordinary Witt defect for a form on an finite and even dimensional space (B_0 carries a nondegenerate symplectic form and is therefore even dimensional). At the second localization, by $rad\,B$, the linear part is replaced by a subspace of finite codimension and the radical is factored out; the space Q is also reduced to the set of forms vanishing on $rad\,B$. As this does not alter the Witt defect of such forms in the finite dimensional case, we let $\omega^B = \omega^{B_0} {}^- Q \cap B^\perp$.

One must check the consistency of such conventions, but this reduces to their correctness in the finite dimensional case, using common localizations.

The initial orientations on the coordinate geometries will not necessarily agree with the ones given here; according to the orientation condition, on a given level of the coordinatization tree, within a given nonorthogonality class, they will be completely correct or completely incorrect. We may change the orientations of the coordinate geometries to agree with our canonical assignment, and nothing is lost. ∎

Example 2.5.8

It is appropriate to return to the canonical unoriented example at this point. Take an unoriented quadratic geometry, and let $\mathcal{M}$ consist of two copies of this geometry, with an identification, and with both possible orientations. To orient this geometry one must name an element of $acl(\emptyset)$.

There are two canonical projectives in this example, with each of the two possible orientations. Our canonical orientation procedure is not available. We can, however, pick an orientation on one of the canonical projective quadratic geometries and extend this orientation to the rest of the structure. Since the orientation is in $acl(\emptyset) - dcl(\emptyset)$, this produces a slightly enriched structure.

If the example is put higher up the coordinatization tree of a structure, it forces us to break the symmetry between elements which are not algebraic over $\emptyset$.

3

Smooth Approximability

3.1 ENVELOPES

We defined standard systems of geometries at the end of the last section. These provide a framework for the construction of Zilber/Lachlan envelopes.

Definition 3.1.1. *Let M be Lie coordinatized.*

1. *A regular expansion of M is the structure obtained by adjoining to M finitely many sorts of M^{eq} with the induced structure.*

   Note that a regular expansion of M is Lie coordinatizable but not literally Lie coordinatized, since the additional sorts are disjoint from the tree structure.

   A regular expansion of M is adequate if it contains a copy of each canonical projective which is nonorthogonal to a coordinatizing geometry of M.

   The remainder of this definition should be applied only to adequate regular expansions of Lie coordinatized structures (as will be seen on inspection of the definition of envelopes, below).

2. *An approximation to a geometry of a given type is a finite or countable dimensional geometry of the same type.*

   This includes, of course, a nondegeneracy condition on the bilinear or quadratic forms involved, and in the case of a quadratic geometry, the quadratic part must be present (a symplectic space with Q empty is not an approximate quadratic space), and ω in the finite dimensional case must actually be the Witt defect.

3. *A dimension function is a function μ defined on equivalence classes of standard systems of geometries, with values isomorphism types of approximations to canonical projective geometries of the given type. (This is actually determined by a dimension, and the type.)*

   By the usual abuse of notation, we construe these functions as functions whose domain consists of all standard systems.

4. *If μ is a dimension function, then a μ-envelope is a subset E satisfying the following three conditions:*

   (i) *E is algebraically closed in M (not M^{eq});*
   (ii) *For $c \in M - E$, there is a standard system of geometries J with domain A and an element $b \in A \cap E$ for which $acl(E, c) \cap J_b$ properly*

contains $acl(E) \cap J_b$;

(iii) *For J a standard system of geometries defined on A and* $b \in A \cap E$,
$J_b \cap E$ *has the isomorphism type given by* $\mu(J)$.

5. *If* μ *is a dimension function and E is a* μ-*envelope we write* $\dim_J(E)$ *for*
$\mu(J)$ *when E meets the domain of J, and otherwise we write* $\dim_J(E) =$
-1; *in the latter case the value* $\mu(J)$ *is irrelevant to the structure of E.*

Our goals are existence, finiteness, and homogeneity of envelopes.

Lemma 3.1.2. *Let* $\mathcal{M}$ *be an adequate regular expansion of a Lie coordinatized structure. Suppose that E is algebraically closed, and satisfies (iii)
with respect to the standard system of geometries J. Suppose that J' is an
equivalent standard system of geometries and that* J, J' *are in* $\mathcal{M}$ *(not just*
$\mathcal{M}^{\mathrm{eq}}$*). Then E satisfies (iii) with respect to* J'.

Proof. We note that as $E \subseteq \mathcal{M}$ it would not make a great deal of sense to
attempt to say something substantial about its intersection with a geometry
lying partly outside $\mathcal{M}$.

Condition (iii) for J' means that if $b' \in E \cap A'$, where A' is the domain of
J', then $E \cap J'_{b'}$ has the structure specified by $\mu(J') = \mu(J)$. The element b'
corresponds to an element b of $E \cap A$, with A the domain of J, and there is a
0-definable bijection between $E \cap J_b$ and $E \cap J'_{b'}$ which is an isomorphism of
weak unoriented structures. This may involve twisting by a field automorphism
or switching the sides of a polar geometry, but does not affect the isomorphism
type. If we use canonical orientations, it will preserve them. ∎

Lemma 3.1.3 (Existence). *Let* $\mathcal{M}$ *be an adequate regular expansion of a Lie
coordinatized structure.*

1. *Let* $E_0 \subseteq \mathcal{M}$ *be algebraically closed in* $\mathcal{M}$ *and suppose that for each
standard system of geometries J with domain A and each* $b \in E_0 \cap A$,
$J_b \cap E_0$ *embeds into a structure of the isomorphism type* $\mu(J)$. *Then* E_0 *is
contained in a* μ-*envelope.*

2. *In particular,* μ-*envelopes exist, for any* μ.

Proof. Let $\mathcal{J}$ be a representative set of standard systems of geometries. By
the previous lemma it suffices to work with $\mathcal{J}$. We may take E containing E_0
maximal algebraically closed such that

$(*)$ For $J \in \mathcal{J}$ with domain A, and $b \in E \cap A$,
$J_b \cap E$ embeds into a structure of the type specified by $\mu(J)$.

We need to check both (ii) and (iii) for E.

We begin with (ii). Suppose $c \in \mathcal{M} - E$. Let $E' = acl(E \cup \{c\})$. Then we
have some $J \in \mathcal{J}$ with domain A, and some $b \in E' \cap A$, for which $J_b \cap E'$
does not embed into a structure of the type specified by $\mu(J)$. If $b \in A \cap E$

then $J_b \cap E$ does embed in such a structure, and (ii) follows. Now suppose that $b \notin A \cap E$. In this case we show that $J_b \cap E = \emptyset \neq J_b \cap E'$, so that (ii) holds also in this case. As E is definably closed it is a subtree of $\mathcal{M}$ with respect to the coordinatizing tree. As b is not definable over E, J_b is orthogonal to the geometries associated with this tree. Thus by induction over this tree, $acl(E) \cap J_b = \emptyset$.

We turn to (iii), and we need only concern ourselves here with $J \in \mathcal{J}$. Suppose that J has domain A, and $b \in E \cap A$, and let B be an extension of $J_b \cap E$ inside J_b of the desired isomorphism type $\mu(J)$. Our claim is that $B \subseteq E$. Let $E' = acl(E \cup B)$. We will argue that E' also has the property $(*)$, so $E' = E$.

If $J' \in \mathcal{J}$ has domain A', and $b' \in A'$, then unless $J' = J$ and $b' = b$, the geometries $J_b, J_{b'}$ are orthogonal and $J_{b'} \cap E' = J_{b'} \cap E$. On the other hand, by Lemma 2.3.3 any element of J_b^{eq} algebraic over E is algebraic over $J_b \cap E$. This applies in particular to any E-definable formula $\varphi(x, y)$ such that for some elements $\mathbf{b} \in B$, $\varphi(x, \mathbf{b})$ isolates an algebraic type over $E \cup B$ in J_b. Thus $J_b \cap E' = J_b \cap acl((E \cap J_b) \cup B) = B$.   ∎

Lemma 3.1.4 (Finiteness). *Let $\mathcal{M}$ be an adequate regular expansion of a Lie coordinatized structure. Suppose that for each standard system of geometries J the dimension function μ is finite. Then every μ-envelope E is finite.*

Proof. E is algebraically closed in $\mathcal{M}$ and hence inherits a coordinatizing tree from $\mathcal{M}$. It suffices, therefore, to check that for any $a \in E$, its successors in the tree form a finite set. We may suppose the successors are of the form $E \cap P_a$ with P_a an a-definable geometry in $\mathcal{M}$, nonorthogonal to some canonical projective geometry J_b with $b \leq a$ in the tree. The size of $J_b \cap E$ is controlled by μ and there is an a-definable bijection between the localization of J_b at $acl(a) \cap J_b$ and the projective version of P_a, so this goes over to E as well. Thus $E \cap P_a$ is finite.   ∎

3.2 HOMOGENEITY

Definition 3.2.1

1. *Let (V, A) be an affine space (a linear space with a regular action) defined over the set B. A is free over B if there is no projective geometry J defined over B for which $A \subseteq acl(B, J)$. An element a, or its type over B, is said to be affinely isolated over B if a belongs to the affine component A of an affine space (V, A) defined and free over B.*

   Note: As a copy of V is definable over A in A^{eq}, it can and will be suppressed in this context.

2. *Let A and A be two affine spaces free over B. They are* almost orthogonal *if there is no pair* $a \in A$, $a' \in A'$ *with* $acl(a, B) = acl(a', B)$.

Lemma 3.2.2 (Uniqueness of Parallel Lines). *Let* (V, A), (V', A') *be almost orthogonal affine spaces defined and free over the algebraically closed set B, with PV and PV' complete 1-types over B. Let J be a projective geometry defined over B, not of quadratic type, and stably embedded in* $\mathcal{M}$. *For* $a \in A$, $a' \in A'$, *and* $c \in J - B$, *the triple* (a, a', c) *is algebraically independent over B.*

Proof. We have (V, A), (V', A'), and J all defined over B. Our definitions amount to the hypothesis that the elements (a, a', c) are pairwise independent over B, so if two of these geometries are orthogonal there is nothing to prove. We suppose therefore that they are all nonorthogonal. In particular, the projectivization PV of V can be identified with part of J.

We consider the structure $J \cup A$. For $a \in A$, A is definable over $J \cup \{a\}$ and hence $J \cup A$ is stably embedded in $\mathcal{M}$. As PV can be identified with part of J, $J \cup A$ carries a modular geometry over B.

Now suppose toward a contradiction that $rk(aa'c/B) = 2$. Take independent conjugates a_1, c_1 of a, c over a'. Then $rk(aca_1c_1/B) = 3$. This takes place in $J \cup A$, so there is $d \in (J \cup A) - B$ algebraic over acB and a_1c_1B, hence in $acl(a', B)$. Thus $acl(dB) = acl(a'B)$ and either $d \in A$, and A, A' are not almost orthogonal, or $d \in J$, and A' is not free over B.   ∎

Lemma 3.2.3. *Let* $\mathcal{M}$ *be Lie coordinatized. Let A be an affine space defined and free over the algebraically closed set B. Let* $B \subseteq B' = acl(B')$ *with B' finite, and let J be a canonical projective geometry associated with A. Assume*

1. *$J \cap B' \subseteq B$;*
2. *$J \cap B$ is nondegenerate (if there is some form or polarity present);*
3. *If J is a quadratic geometry, then its quadratic part Q meets B.*

   Then A either meets B', or is free over it.

Proof. We remark that if A does not meet B', A need not remain a geometry over B', but will split into a finite number of affine pregeometries over B'. We will call A free over B' if this applies to each of the associated geometries over B'.

The claim will be proved by induction with respect to the coordinatization of the algebraically closed set B' relative to B, inherited from $\mathcal{M}$. Accordingly we may take $B' = acl(B, a')$, where a' comes from an affine, quadratic, or projective geometry A' defined over B.

Assume that $A \cap B' = \emptyset$ and some affine part A_0 of A relative to B' is contained in $acl(B, a', J')$ with $J' = J_{b'}$ projective and defined over B'. As $J' \subseteq acl(J, b')$ the same applies with J' replaced by J, that is: $A_0 \subseteq acl(B, a', J)$, while $A \cap acl(B, J) = \emptyset$. It follows that A' and J are nonorthogonal, and that $A' \cap acl(B, J) = \emptyset$. In view of (iii) we have A' affine, and easily free over B.

If A and A' are not almost orthogonal over B, then B' meets A. Suppose therefore that A and A' are almost orthogonal over B. Then we will apply the previous lemma. Choose $a \in A$. As $a \in acl(B, a', J)$, and the geometry of (A, J) is modular, there is $c \in J \cap acl(Baa')$ with $a \in acl(Ba'c)$. Then $c \notin B$, and in view of (iii) we may suppose c is not in the quadratic part of J, if there is a quadratic part.

Let J_B be the localization of J over B. By hypothesis (iii) this is not a quadratic geometry. By hypothesis (ii) J is in the algebraic closure of $B \cup J_B$; normally over B, J would break up into a number of pregeometries, at least one $((J \cap B)^\perp)$ sitting over the localization, while some of the cosets would be affine pregeometries. However, since $J \cap B$ is nondegenerate, all elements of J lie in translations by elements of B of $(J \cap B)^\perp$. Of course, when forms are absent, the situation is trivial.

Replacing c momentarily by an element of J_B having the same algebraic closure over B, we may apply the previous lemma to a, a', c, reaching a contradiction.   ∎

Lemma 3.2.4. *Let $\mathcal{M}$ be an adequate regular explansion of a Lie coordinatized structure, let μ be a dimension function, and let E and E' be μ-envelopes. If $A \subseteq E$, $A' \subseteq E'$ are finite, and $f : A \to A'$ is elementary in $\mathcal{M}$, then f extends to an elementary map from E to E'. In particular, μ-envelopes are unique, and (taking $E = E'$) homogeneous.*

Proof. It suffices to treat the case in which E and E' are finite, as the existence and finiteness properties then suffice for a back-and-forth argument using finite envelopes. What we must show is that if $A \neq E$ then there is an extension of f to $acl(A\cup\{b\})$ for some $b \in E - A$. There are essentially two cases, depending on whether we are trying to add a point to the domain coming from a canonical projective geometry, or we are extending to the other points of the envelope. We may suppose A and A' are algebraically closed.

Case 1. There is a standard system of geometries J and an $a \in A$ for which $J_a \cap E$ is not contained in A.

Expand J_a to a basic projective geometry $J_{a^*}^\circ$ defined over $a^* = acl(a)$. Let L, L' be finite dimensional linear geometries covering $J_{a^*}^\circ \cap E$ and $J_{fa^*}^\circ \cap E'$, respectively. Then L and L' are isomorphic, and their isomorphism type is characterized by its type, dimension, and Witt defect (if applicable).

As f is elementary, it gives a partial isomorphism between some $J_a \cap E$ and $J_{fa} \cap E'$, which lifts to an elementary map between the corresponding parts of L and L'. Let $\hat{f}$ be an extension of f by an isomorphism of L with L'. The existence of such a compatible extension is trivial in the absence of forms and given by Witt's theorem [Wi] otherwise, with the exception of the polar and quadratic cases. The polar case is quite straightforward. In the quadratic case one first extends f so that its domain meets Q, and then the problem reduces to the orthogonal case, in other words to Witt's theorem.

By weak elimination of imaginaries and stable embedding, since $A = acl\, A$, we find that $tp(A/L \cap A)$ determines $tp(A/L)$. Similarly, the type $tp(A'/L' \cap A')$ determines $tp(A'/L')$. Implicit in this determination is knowledge of the type of L or L' over $\emptyset$. Since $\hat{f}$ preserves the two relevant types, it preserves $tp(A/L)$ and is thus elementary.

Case 2. For any standard system of geometries J, and any $a \in A$, $J_a \cap E \subseteq A$.

It follows that the same applies to A'. We extend f to a minimal element a in the coordinatization tree for E, not already in the domain. So the tree predecessor b of a is in A, and a is not algebraic over b. Accordingly a belongs to a geometry J_b which is nonorthogonal to a canonical projective geometry. As we are not in Case 1, J_b is affine, and free over A. If f is extended to $acl(A) \cap \mathcal{M}^{\text{eq}}$ we may take J_b basic.

In E' we have, correspondingly, J_{fb} affine and free over A'. However, as E' is an envelope, the maximality condition (clause (ii)) implies that J_{fb} cannot be free over E'. Lemma 3.2.3 applies in this situation, to the affine space J_{fb} and the algebraically closed sets A' and E', in view of the hypothesis for Case 2. Thus the conclusion is that J_{fb} meets E'.

We will next find an element a' of $J_{fb} \cap E'$ satisfying the condition

$$(a, \lambda) = (a', f\lambda) \text{ for all } \lambda \in J_b^* \cap A \text{ (the affine dual)}.$$

Here one should, strictly speaking, again extend f to the algebraic closure of A in $\mathcal{M}^{\text{eq}}$. We consider a stably embedded canonical projective geometry P associated with J_b. Then P is b-definable and the projectivization of the linear space V_b which acts regularly on J_b is definably isomorphic to one of the sorts of the localization P/b of P at b. By our case assumption $P \cap E$ is as specified by μ and is, in particular, nondegenerate. The same applies to $P' \cap E'$. Thus the action of the definable linear dual of $V_{b'}$ is represented, in its action on $V_{b'} \cap E'$, by elements of A' (or $acl(A') \cap \mathcal{M}^{\text{eq}}$, more precisely). As E' meets the affine space $J_{b'}$, the same applies to the affine dual. Again by the linear nondegeneracy and the fact that E' meets $J_{b'}$, the specified values for $(a', f\lambda)$ can be realized in $E' \cap J_{b'}$. We extend f by $f(a) = a'$.

Now the type of A over (PV_b, J_b, J_b^*) is determined by its type over its algebraic closure in that geometry, and this applies in particular to the type of A over a. So in order to see that f remains elementary, it suffices to check that a and a' have corresponding types over $A^{\text{eq}} \cap (PV_b, J_b, J_b^*)$ and its f-image; and this is what we have done. ∎

Corollary 3.2.5. *Let $\mathcal{M}$ be an adequate regular expansion of a Lie coordinatized structure. Then a subset E of $\mathcal{M}$ is an envelope if and only if the* *following conditions are satisfied:*

1. *E is algebraically closed;*
2. *For any $c \in M - E$, there is a projective geometry J defined over E, not quadratic, and an element $c' \in (J \cap acl(Ec)) - E$;*
3. *If c_1, $c_2 \in E$ are conjugate in $\mathcal{M}$ and $D(c_1)$, $D(c_2)$ are corresponding conjugate definable geometries, then $D(c_1) \cap E$ and $D(c_2) \cap E$ are isomorphic.*

This does not depend on a particular coordinatization of $\mathcal{M}$.

3.3 FINITE STRUCTURES

In this part we summarize some useful facts applying to finite geometries and their automorphism groups, notably the result of [KLM].

Definition 3.3.1. *A simple Lie geometry L is either a weak linear geometry of any type other than polar or quadratic, the projectivization of such a geometry, or the affine or quadratic part of a geometry; in the latter case the "missing," linear part is to be considered as encoded into L^{eq}.*

These do not have the best properties model theoretically, and a polar geometry cannot be recovered at all from a single simple Lie geometry, but apart from this, at the level of C^{eq} there is little difference between simple Lie geometries and the geometries considered previously.

Definition 3.3.2

1. *A coordinatizing structure of type (e, K) and dimension d is a structure C with a transitive automorphism group, carrying an equivalence relation E with $e < \infty$ classes, such that each class carries the structure of a simple Lie geometry over the finite field K, of dimension d. (One could include the type of the geometry as well in the type of C.)*

2. *Let C be a coordinatizing structure of type (e, K) and dimension d, and let τ be the type over the empty set of a finite algebraically closed subset (not sequence) t of C. The Grassmannian $\Gamma(C, \tau)$ is the set of realizations of the type τ in C, with the structure induced by C. It is said to have type (e, K, τ) and dimension d.*

3. *Let C be a coordinatizing structure. C is proper if each equivalence class of C as a geometry is canonically embedded in C, or equivalently if the automorphism group induced on each class is dense in its automorphism group as a geometry (in the finite dimensional case, dense means equal). If C is finite dimensional, it is semiproper if the automorphism group of C induces a subgroup of the automorphism group G of the geometry which contains $G^{(\infty)}$.*

4. *A structure is* primitive *if it has no nontrivial 0-definable equivalence relation.*

Fact 3.3.3 [KLM]. *For each k there is n_k such that for any finite primitive structure $\mathcal{M}$ of order at least n_k, if $\mathcal{M}$ has at most k 5-types then $\mathcal{M}$ is isomorphic to a semiproper Grassmannian of type (e, K, τ) with $e, |K|, |\tau| \leq k$, where $|\tau|$ has the obvious meaning.*

As noted in the introduction, D. Macpherson found [Mp2] that the method of proof of [KLM] suffices to prove the same fact with 5 reduced to 4. The statement is quite false for 3.

The next set of facts is standard in content, though not normally phrased precisely as follows.

Fact 3.3.4 [CaL]

1. *Let k be an integer. There is a $d = d_k$ such that for any finite basic simple projective Lie geometry L of dimension at least d we have*

   (i) *The socle G of $Aut(L)$, is simple and nonabelian, and $Aut(L)/G$ is solvable of class at most 2;*

   (ii) *G and $Aut\,L$ have the same orbits on L^k;*

   (iii) *The automorphism group of L as a weak* geometry *coincides with $Aut\,G$. with one exception: if L is a pure vector space then the automorphism group of L is a subgroup of index 2 in $Aut\,G$, and the full group $Aut\,G$ is realized geometrically as the automorphism group of the weak polar geometry (L, L^*).*

2. *If J_1, J_2 are nondegenerate basic projective geometries, not quadratic, of large enough dimension, and their automorphism groups have isomorphic socles, then they are isomorphic as weak geometries.*

Here our policy of leaving the degenerate case to fend for itself may be too lax; but the statement certainly applies also in the context of $Sym(n)$ and $Aut(n)$.

Remark 3.3.5. Note that the automorphism groups of the basic geometries are classical groups with no Galois action. In the first statement we ignore 4-dimensional symplectic groups over fields of characteristic 2 and 8-dimensional orthogonal groups of positive Witt defect by taking $d > 8$. The polar geometry implements a "graph automorphism," of the general linear group in any dimension. The graph automorphism of order 2 for Chevalley groups of type D_n is part of the geometric automorphism group. G is usually equal to the commutator subgroup of $Aut\,L$, with exceptions in the orthogonal case (and a few small exceptions that can be ignored here).

Fact 3.3.6. *For any finite basic simple linear geometry V of dimension at least 5, if $G = (Aut\,V)^{(\infty)}$ acts on an affine space A over V so as to induce its* *standard action on V, then either G fixes a point of A or the characteristic is 2, G is the symplectic group operating on its natural module V, and the action of G on A is definably equivalent to its action on Q, the space of quadratic maps on V associated to the given form.*

Proof. Taking any point $a \in A$ as a base point, the function $f(g) = a^g - a$ can be construed as a function from G to V, and is a 1-cocycle. Change of base point gives a cohomologous cocycle. If this cocycle is trivial, it means we may choose the base point so that this cocycle vanishes, and a is a fixed point for the action of G.

Typically the first cohomology group for a (possibly twisted) Chevalley group on its natural module vanishes; see the tables in [JP], for example. Some rather large counterexamples are associated with exceptional Chevalley groups, but for the classical types ($A - D$, possibly twisted) restricted to dimension greater than 4, the only counterexamples involving natural modules are 1-dimensional cohomology groups for symplectic groups in characteristic 2 (these are listed twice in [JP], once as C_n and once as B_n, since the natural module for the odd dimensional orthogonal groups in even characteristic corresponds to a representation of this group as the symplectic group in one lower dimension). This is the case in which we have Q, or more exactly αQ for $\alpha \in K^\times$. The latter can be thought of most naturally as the space of quadratic forms inducing $\alpha\beta$, where β is the given symplectic form on V, but can also be viewed as the space Q with the action $q \mapsto q + \lambda_v^2$ replaced by the action $q \mapsto q + \lambda_{\alpha^{1/2}v}^2$.

Thus we can either consider A as isomorphic to Q, by an isomorphism which is not the identity on V, or as definably equivalent to Q over V, holding V fixed and rescaling the regular action on A; our formulation of the result reflects the second alternative.   ∎

Remark 3.3.7. It seems advisable to remember that the "Q" alternative in the preceding statement is in fact αQ for some unique $\alpha \in K$.

Fact 3.3.8 [CaK]. *Let G be a subgroup of a classical group acting naturally on a finite basic simple classical projective geometry P, and suppose that G has the same action on P^3 as Aut P. Then G contains $(\text{Aut } P)^{(\infty)}$ (the iterated derived group).*

This iterated derived group is at worst $(\text{Aut } P)^{(2)}$ and is a simple normal subgroup with solvable quotient.

Remark 3.3.9. In this connection, our general policy of leaving the degenerate case to fend for itself is definitely too lax. A similar statement does apply also in the context of $Sym(n)$ and $Aut(n)$, with 6-tuples in place of 3-tuples, but one needs the classification of the finite simple groups to see this.

Fact 3.3.8 is phrased rather differently in [CaK], as the result is considerably sharper in more than one respect. Here we ignore low dimensional examples and also invoke a significantly stronger transitivity hypothesis. A somewhat more complete statement of the result of [CaK] goes as follows.

Fact 3.3.10 [CaK, cf. Theorem IV]. *Let* $G \leq \Gamma L(n, q)$, $n \geq 3$, *and suppose* G *is 2-transitive on the corresponding projective space. Then either* $G \geq SL(n, q)$ *or* $G \leq SL(4, 2)$.

Fact 3.3.11 [CaK, cf. Theorem IV]. *Let* $G \leq H = \Gamma Sp(n, q)$, $\Gamma O^\epsilon(n, q)$, *or* $\Gamma U(n, q)$ *with* $n > 13$ *and suppose that* G *has the same orbits on 2-dimensional spaces as* H. *Then* $G \geq H^{(\infty)}$.

Theorem IV of [CaK] varies from Fact 3.3.10 in the following respects. The transitivity hypothesis is weaker, amounting to transitivity on pairs consisting of two isotropic nonorthogonal lines. This allows three low dimensional exceptions and two families defined over the field F_2, where G normalizes a classical subgroup with coefficients in F_4, so that G has more than one orbit on totally isotropic planes.

Lemma 3.3.12. *Let* H *be a normal subgroup of a product*

$$G = G_1 \times \cdots \times G_n$$

such that H *projects surjectively onto each product of the form* $G_i \times G_j$. *Then* G/H *is nilpotent of class at most* $n - 2$. *In particular, if* G *is perfect then* $G = H$.

Proof. Let σ_i for $1 \leq i \leq n - 1$ be a sequence of elements of G_n and for each i let $\sigma_i^* \in G$ be an element of H which projects onto σ_i in the nth coordinate, and 1 in the i-th coordinate. Then any iterated commutator $\gamma(\sigma_i^*)$ in the elements σ_i^* will project onto $\gamma(\sigma_i)$ in G_n, and 1 in the other coordinates. It follows easily that any iterated commutator of length $n - 1$ belongs to H, and our claim follows. ∎

Remarks 3.3.13.
The proof of Fact 3.3.6 actually involves a great deal of calculation, somewhat disguised by the fact that the reference [JP] presents the final outcome in tabular form. A qualitative version of this, sufficient for our purposes, can be obtained by postponing the issue somewhat and making use of our later results. We will indicate this approach.

View (A, V) as a structure by endowing it with all invariant relations. Replacing the bound "5" by "sufficiently large," we may take V to have a nonstandard dimension. If we show that A has either a 0-definable point, or quadratic structure, then the same follows for sufficiently large finite dimensions.

The induced structure on V is that of a standard linear geometry. Let V' be the structure induced on V by (V, A, a) with a a point of A. Note that V'

interprets the triple (V, A, a). One cannot expect V' to be stably embedded, in view of the characteristic 2 case, but we still expect

$(*)$ V' is Lie coordinatized.

Given $(*)$, one deduces Fact 3.3.6 from the theorem on reducts and the recognition lemmas: by Proposition 7.5.4 (V, A) is weakly Lie coordinatized. By Lemma 6.2.11 V is part of a basic linear geometry in this structure, and Proposition 7.1.7 recognizes A.

The theorem on reducts can also be used in the proof of $(*)$ itself. Note that any two unstable linear geometries interpret each other, provided only that the characteristics of the base fields are equal. Once reducts are under control, one can expand the geometry to a polar geometry over a field of size greater than 2. This has the effect of reducing all cases of $(*)$ to the simplest case of Fact 3.3.6, namely the general linear group, which can be handled by a direct argument.

3.4 ORTHOGONALITY REVISITED

For simplicity we will work for some time in a nonstandard extension of the set theoretic universe in which we have infinite integers. This gives a rigorous basis for the treatment of sequences of finite structures of increasing size in terms of one infinitely large structure of integral cardinality. In this context it will be important to distinguish internal and external objects, notably in connection with the languages used, and the supply of automorphisms available.

Definition 3.4.1. *Let $\mathcal{M}$ be an internally finite structure with internal language L_0 in a nonstandard extension of the universe of set theory. Then $\mathcal{M}^*$ is the structure with the same universe, in a language whose atomic relation symbols consist of names for all the relations in finitely many variables defined on $\mathcal{M}$ by L_0-formulas.*

Observe that $\mathcal{M}^*$ is not an element of the nonstandard universe. If $\mathcal{M}$ is a nonstandard finite model of a standard theory T in the language L, then the corresponding language L_0 (normally called L^* in this case) is the language corresponding to L in the nonstandard universe; this has more variables than L (x_n for all integers n, standard or nonstandard), and more importantly, consists of arbitrary internally finite well-formed formulas in its language. This includes formulas with infinitely many (but internally finitely many) free variables; these are discarded in forming the language for $\mathcal{M}^*$, so $\mathcal{M}^*$ is a reduct of $\mathcal{M}$ from the nonstandard language L^*, one which is in general richer than the reduct of $\mathcal{M}$ to the standard language L. For a concrete example, consider a discrete linear order of nonstandard finite length: among the predicates of $\mathcal{M}^*$ one has, for example, the distance predicates $D_n(x, y)$ in two variables, for every n up to the (nonstandard) size of the order. Of course, in this case there are no nontrivial internal automorphisms of $\mathcal{M}$; in fact, there are no nontrivial automorphisms of $\mathcal{M}^*$.

Lemma 3.4.2. *Let $\mathcal{M}$ be an internally finite structure, and J a finite disjoint union of basic 0-definable projective simple Lie geometries with no additional structure. Let G be Aut J and let G_1 be $(\text{Aut } J)^{(\infty)}$ (the iterated derived group), where both Aut J and Aut $J^{(\infty)}$ are understood internally (the latter coinciding with the internal socle here), and automorphisms are taken with respect to the geometric structure. Let H be the group of automorphisms of J which are induced by internal automorphisms of $\mathcal{M}$. Then J is canonically embedded in $\mathcal{M}^*$ if and only if H contains G_1.*

Proof. Suppose first that J consists of a single projective geometry. J is canonically embedded in $\mathcal{M}^*$ if and only if for each finite n, G and H have the same orbits on n-tuples in J; applying Fact 3.3.4, part 1(ii), this means that G_1 and H have the same orbits on n-tuples in J. This certainly holds if $H \supseteq G_1$. Conversely if H has the same orbits on J as G, it contains G_1 by Fact 3.3.8.

The argument is similar in the general case, but we must justify further the claim that if H acts on n-tuples of J as does G, then it contains G_1. Arguing inductively, it suffices to show that the pointwise stabilizer of J_1 in H acts on m-tuples from $J_2 \times \cdots \times J_n$ as G_1 does. Let $B \subseteq J_2 \times \cdots \times J_n$ have cardinality m, and let $g \in G_1$. By the argument of the first part, the action of the pointwise stabilizer H_{B^g} on J_1 induces the action of g on J_1. Hence in its action on $J_2 \times \cdots \times J_n$, H_{J_1} has the same orbits on m-tuples as G; by induction then, H_{J_1} induces the action of G_1 on $J_2 \times \cdots \times J_n$. It follows that H induces G_1. ∎

Lemma 3.4.3. *Let $\mathcal{M}$ be an internally finite infinite structure. Let J_1, J_2 be a pair of basic pure projective geometries (with no forms) defined and orthogonal over the algebraically closed set A in the sense that $(J_1, J_2; J_1 \cap A, J_2 \cap A)$ is canonically embedded in $\mathcal{M}$. Let $J = J_1 \cup J_2$, $A_J = A \cap J$. Then the permutation group G induced on J by the internal automorphism group of $\mathcal{M}$ contains $\text{Aut}(J; A_J)^{(\infty)}$ (which in this case is just the commutator subgroup of $\text{Aut}(J; A_J)$). All group theoretic notions are to be understood internally here.*

Proof. For notational definiteness let us assume that $A \cap J_i$ is nonempty for each i. In the linear model we have vector spaces V_i with $PV_i = J_i$ and we will take $U_i = acl(A) \cap V_i$, and decompose $V_i = U_i \oplus W_i$. Then we may check

$$Aut(J_i; A \cap J_i) \simeq Hom(W_i, U_i) \rtimes GL(W_i).$$

Our claim is that the group G contains the product $Hom(W_i, U_i) \rtimes SL(W_i)$, acting on J. We know that on the localizations $Aut\,\mathcal{M}$ induces $PSL(W_1) \times PSL(W_2)$ as these geometries are orthogonal. Let H_1 be the kernel of the natural map from G to $Aut\,J_2/(A \cap J_2)$. Then H_1 covers at least $PSL(W_1)$ and is normal in G. It follows that the same applies to the perfect subgroup $H_1^{(\infty)}$. Now $H_1^{(\infty)}$ projects trivially into the second factor and may therefore be thought of as a normal subgroup of $Aut(J_1; A \cap J_1)$ covering $PSL(W_1)$; any such subgroup contains $Hom(W_1, A \cap J_1) \rtimes SL(W_1)$, by inspection.   ∎

Remark 3.4.4. We are working here with automorphisms of *pointed* projective geometries, in which constants have been added. It is not always possible to reduce their analysis to a localization. In a similar vein, Lemma 3.4.2 may be proved for pointed pure projective geometries as well, or for that matter for any pointed projective geometries, if we are willing to write out the stabilizers of various sets.

Definition 3.4.5. *A collection of A-definable sets S_i is said to be* jointly orthogonal *over A in $\mathcal{M}$ if the disjoint union of the structures $(S_i, acl(A) \cap S_i)$ is canonically embedded in $\mathcal{M}$.*

Lemma 3.4.6. *Let J_i be defined over A in $\mathcal{M}$, with weak elimination of imaginaries, and let $B \subseteq J = \bigcup_i J_i$. Then the J_i are jointly orthogonal in $\mathcal{M}$ over A if and only if they are jointly orthogonal in $\mathcal{M}$ over $A \cup B$.*

Proof. If they are jointly orthogonal over A and R is a relation on J definable from $A \cup \bigcup_i acl(AB) \cap J_i$, then R is the specialization of a 0-definable relation S over J to parameters from $\bigcup_i acl(AB) \cap J_i$. Accordingly S is a boolean combination of products of $(acl(A) \cap J_i)$-definable relations on J_i, and after specialization the same applies to R over AB.

Conversely, assuming orthogonality over $A \cup B$, let R be A-definable on J. This is definable by hypothesis in J, with respect to parameters from $\bigcup_i acl(AB) \cap J_i$. Viewing R as an element of J^{eq}, let $e = acl(R) \cap J$. By weak elimination of imaginaries, R is e-definable and $e \subseteq acl(A) \cap J$.   ∎

Lemma 3.4.7. *Let $\mathcal{M}$ be an internally finite structure. Let J_i $(i \in I)$ be canonically embedded projective Lie geometries in $\mathcal{M}^*$, defined over, and orthogonal in pairs over, the set A in $\mathcal{M}^*$. Then they are jointly orthogonal over A in $\mathcal{M}^*$.*

Proof. Let $A_i = acl(A) \cap J_i$. The assumption is that $(J_i \cup J_j; A_i \cup A_j)$ is canonically embedded in $\mathcal{M}^*$. Extend A by finite subsets B_i of J_i containing A_i so that B_i is a nondegenerate subspace containing a quadratic point, if possible. In the pure projective case $B_i = A_i$. We may replace A by $B = A \cup \bigcup_i B_i$. Then A_i is replaced by B_i, the geometries continue to be pairwise orthogonal, and it suffices to prove joint orthogonality over AB. For this, by the choice of B_i, except in the pure projective case it suffices to go to the (nondegenerate) localizations, which are definably equivalent over B_i to the previous structures. Now we consider the group H of permutations induced by $Aut \mathcal{M}$ on $\bigcup_i (J_i; B_i)$. Write G_i for $Aut(J_i; B_i)^{(\infty)}$. Applying Lemma 3.4.2 of §3.3 to $H^{(\infty)}$, using the pairwise orthogonality, we find $H \supseteq \prod_i G_i$. By Lemma 3.4.2 and the remark following Lemma 3.4.3 (used in the more straightforward of the two directions) our claim follows. ∎

Lemma 3.4.8. *Let $\mathcal{M}$ be an internally finite structure. Let J_1, J_2 be 0-definable basic simple projective Lie geometries canonically embedded in $\mathcal{M}^*$. Then in $\mathcal{M}^*$ we have one of the following:*

1. *J_1 and J_2 are orthogonal.*
2. *There is a 0-definable bijection between J_1 and J_2.*
3. *J_1 and J_2 are of pure projective type—that is, with no forms—and there is a 0-definable duality between them making the pair (J_1, J_2) a polar space.*

Proof. Let S be the internal permutation group induced on $J = J_1 \cup J_2$ by internal automorphisms of $\mathcal{M}$ and let G_i be the internal automorphism group of the geometry J_i. Set $S_1 = S \cap (G_1 \times G_2)^{(\infty)}$, again working internally (as we will throughout). As S projects onto G_i, $S^{(\infty)} \subseteq S_1$ projects onto $G_i^{(\infty)}$ for $i = 1, 2$. As $G_i^{(\infty)}$ is simple, $S^{(\infty)}$ is either the full product or the graph of an isomorphism between $G_1^{(\infty)}$ and $G_2^{(\infty)}$.

In the first case J_1 and J_2 are orthogonal by Lemma 3.4.2. In the second case, by Fact 3.3.11, the geometries J_1 and J_2 are isomorphic as weak geometries, and if we identify them by an isomorphism, thereby identifying their automorphism groups, S_1 is then the graph of an automorphism. With the exception of the pure projective case, this automorphism is an inner automorphism with respect to the full automorphism group of the geometry, by Fact 3.3.4, 1(iii); in the exceptional case it may be the composition of an inner automorphism and a graph automorphism. If S_1 is the graph of an inner automorphism corresponding to an isomorphism $h : J_1 \simeq J_2$, then as S_1 is normal in S, this isomorphism is S-invariant, hence 0-definable. In the exceptional case S_1 can be viewed as an isomorphism of J_1^* and J_2; in particular, J_1^* is interpretable in $\mathcal{M}$, and is 0-definably isomorphic with J_2. ∎

Lemma 3.4.9. *Let $\mathcal{M}$ be an internally finite structure. Let A be a 0-definable basic affine space, with corresponding linear and projective geometries V and J. Suppose that J is canonically embedded in $\mathcal{M}^*$. Then one of the following holds in $\mathcal{M}^*$:*

1. *A is canonically embedded in $\mathcal{M}^*$.*
2. *There is a 0-definable point of A in $\mathcal{M}^*$.*

3. *J is of quadratic type and there is a 0-definable bijection of A with αQ for some unique α.*

Proof. As usual all computations with automorphisms will be taken relative to the internal automorphism groups.

We argue first that V is canonically embedded in $\mathcal{M}^*$. Let V_1 be V with all 0-definable relations from $\mathcal{M}$. Then J is canonically embedded in (J, V_1), and stably embedded since $V_1 \subseteq acl(J)$. For $a \in V$, $V_1 \subseteq dcl(Ja)$, and hence $(V_1, a) = (V, a)$ as structures. By weak elimination of imaginaries for V, it follows that $V_1 = V$ as structures.

Now consider

$$U = \{v \in V : \text{Translation by } v \text{ is an automorphism of } A \text{ over } V\}.$$

For v in U let τ_v be the corresponding translation map on A. Then for $\alpha \in Aut\,\mathcal{M}^*$ we have $\tau_v^\alpha = \tau_{\alpha^{-1}v}$. Thus U is $(Aut\,\mathcal{M}^*)$-invariant, and hence also 0-definable in $\mathcal{M}^*$, since $\mathcal{M}$ is internally finite. But V is canonically embedded in $\mathcal{M}^*$, so $U = V$ or $U = (0)$.

If $U = V$ then A is canonically embedded in $\mathcal{M}^*$, since V is. Suppose that $U = (0)$. Every automorphism of V extends to $\mathcal{M}^*$ and hence to A; as $U = (0)$, this extension is unique, and $Aut\,V$ acts on A. By Fact 3.3.6, we have either a fixed point or a bijection with αQ, as in possibilities (2,3) above, fixed by $(Aut\,V)^{(\infty)}$. Furthermore, the fixed point or bijection, as the case may be, is unique, as otherwise this $(Aut\,V)^{(\infty)}$ would fix correspondingly either a point of V, or a nonidentity bijection of αQ with itself. The first alternative is obviously impossible. In the second case, if $q \in \alpha Q$ is moved by the bijection, say $q \mapsto q + \alpha \lambda_v^2$, then v is fixed by the corresponding orthogonal group, which is again a contradiction. Thus the unique fixed point, or the unique bijection with αQ, is fixed by $Aut\,\mathcal{M}^*$.   ∎

3.5 LIE COORDINATIZATION

In this section we introduce the notion of a locally Lie coordinatized structure, which is approximately a structure coordinatized in the manner of [KLM] (in other words, without concern for stable embedding), and we check that the internally finite structures associated with 4-quasifinite structures are biinterpretable with locally Lie coordinatized structures, which is another way of phrasing the results of [KLM] (with 5 reduced to 4). Then to complete the proof of the equivalence of the first five conditions given in Theorem 2, we show that 4-quasifinite locally Lie coordinatized structures are Lie coordinatizable. See the discussion at the end of the present section for a review of the situation up to this point.

Definition 3.5.1. *A structure $\mathcal{M}$ in some nonstandard set theoretic universe is* locally Lie coordinatized *if it has nonstandard finite order, has finitely many 1-types, carries a tree structure of finite height whose unique root is 0-definable, and has a collection $\mathcal{J}$ of pairs (b, J) with $b \in \mathcal{M}$ and $J \subseteq \mathcal{M}$ a b-definable component of a b-definable basic semiprojective, linear, or affine geometry, satisfying the following conditions:*

1. *If a is not the root, then there is $b < a$ such that either $a \in acl(b)$ or there is a pair $(b, J_b) \in \mathcal{J}$ with $a \in J_b$.*
2. *If $(b, J) \in \mathcal{J}$ with J semiprojective or linear then J is canonically embedded in $\mathcal{M}$.*
3. *Affine spaces are preceded in the tree by their linear versions.*

Lemma 3.5.2. *Let Γ be an infinite dimensional proper Grassmannian of type (e, K, τ), and $a \in \Gamma$. Then there are elements $a_0, \ldots, a_n \in \Gamma^{eq} \cap acl(a)$ and Lie geometries J_i, possibly affine, with J_i 0-definable and canonically embedded relative to the structure $(\Gamma; a_0, \ldots, a_i)$, such that $a_0 \in acl(\emptyset)$, $a_{i+1} \in J_i$, and $a \in acl(a_0, \ldots, a_n)$.*

Proof. The components J of the underlying coordinatizing structure $\mathcal{C}$ can be recovered from equivalence relations on pairs from Γ. Let a_0 consist of these components as elements of Γ^{eq}, together with enough elements of $acl(\emptyset)$ in $\mathcal{C}^{eq}$ to make them all basic. We define a_i inductively, stopping when $a \in acl(a_0, \ldots, a_i)$. Given $(a_0, \ldots, a_i)$, with a not algebraic over them, pick a component J meeting $acl(a) - acl(a_0, \ldots, a_i)$ and let a' be a point of the intersection. Consider the localization $\bar{J} = J/(a_0, \ldots, a_i)$. This is not in general the full quotient of J modulo algebraic closure relative to $(a_0, \ldots, a_i)$, but just a part of that. The remainder consists of various geometries which are either 0-definably equivalent to the localization, or affine over it. In particular, we may take a' to represent either an element of this localization or an element of an affine geometry over the localization. More precisely, there is an element a'' lying either in the localization $\bar{J}$, or in an affine geometry over it, for which $acl(a_0, \ldots, a_i, a') = acl(a_0, \ldots, a_i, a'')$. We set $a_{i+1} = a''$ and correspondingly $J_i = \bar{J}$ or an affine geometry over $\bar{J}$.

The localizations are canonically embedded in $(\Gamma; a_0, \ldots, a_i)$. In the affine case Lemma 3.4.9 applies. If the affine space is actually a copy of Q, then a'' is taken in Q (which is part of the semiprojective model). ∎

Lemma 3.5.3. *Let $\mathcal{M}$ be a structure, k an integer, and let Ψ be a finite set of first order formulas in four free variables. Suppose that for every first order sentence φ true in $\mathcal{M}$ there is a finite model $\mathcal{M}'$ satisfying*

1. *$\mathcal{M}' \models \varphi$.*
2. *$\mathcal{M}'$ has at most k 4-types.*
3. *Every 0-definable 4-ary relation on $\mathcal{M}'$ is defined by one of the formulas*

in Ψ.

Then $\mathcal{M}$ is bi-interpretable with a locally Lie coordinatized structure $\mathcal{M}'$
which forms a finite cover of $\mathcal{M}$: $\mathcal{M}'$ has $\mathcal{M}$ as a 0-definable quotient with
finite fibers (see §4.5 for a formal discussion of covers).

Proof. These conditions imply that $\mathcal{M}$ itself has at most k 4-types, and that
every 4-ary relation on $\mathcal{M}$ is defined by one of the formulas in Ψ. In particular,
one can select a maximal chain $E_0 < \ldots < E_d$ of 0-definable equivalence
relations on $\mathcal{M}$ and we may suppose that in all the models $\mathcal{M}'$ this chain
remains a maximal chain of 0-definable equivalence relations (making use,
among other things, of condition (1)). We take $E_i < E_{i+1}$ to mean that E_{i+1}
is coarser than E_i.

For i fixed, and $a \in \mathcal{M}$, we consider the E_{i+1}-class C of a, and its quotient
C/E_i. It will suffice to prove that C/E_i is either finite or a proper Grassmannian, as we can then coordinatize $\mathcal{M}$ by coordinatizing each infinite section
C/E_i, starting from the coarsest, using Lemma 3.5.2; of course, if C/E_i is finite, then its elements are algebraic over C. When projective geometries occur
they can be replaced by semiprojective ones in $\mathcal{M}^{\mathrm{eq}}$.

If C/E_i is infinite, then by [KLM], specifically by Fact 3.3.3, above, we
may suppose that in the finite structures $\mathcal{M}'$ approximating $\mathcal{M}$ in the sense
of clauses (1–3) above, the corresponding set C'/E_i' carries the structure of a
semiproper Grassmannian of fixed type. There are 4-place relations R_i which
encode the components of the coordinatizing structure underlying the Grassmannian, as well as the geometric structure on this coordinatizing structure.
Primarily, the R_i should be equivalence relations on pairs, so as to encode the
elements of the coordinatizing structure; one can also encode, e.g., a ternary
addition relation, with some care, by using four variables in the Grassmannian.

There is also a statement $\gamma(R_1, \ldots, R_n)$ expressing the fact that C'/E_i is a
Grassmannian of the given type for this coordinatizing structure. Accordingly
in view of our hypotheses, a formula of the same type will apply to C/E_i,
for some choice of the R_i, and C/E_i is the Grassmannian of a coordinatizing
structure.

To conclude we must check properness: that is, in C/E_i, we claim that each
0-definable relation S is geometrically definable (i.e., definable from the structure with which the Grassmannian inherits from the coordinatizing structure)
over $acl(\emptyset)$. For fixed S this will hold in sufficiently large finite approximations $\mathcal{M}'$ and by (1) this property passes to $\mathcal{M}$.   ∎

Corollary 3.5.4. *If $\mathcal{M}$ is strongly 4-quasifinite, then $\mathcal{M}$ is bi-interpretable*
with a locally Lie coordinatized structure which forms a finite cover of $\mathcal{M}$.

Lemma 3.5.5. *Let $\mathcal{M}$ be an internally finite structure and suppose that $\mathcal{M}^*$*
has a finite number k of 4-types. Then $\mathcal{M}^*$ is bi-interpretable with a locally
Lie coordinatized structure which forms a finite cover of $\mathcal{M}^*$.

Proof. We apply the previous lemma. Let Ψ be a set of representatives for the internally 0-definable formulas in 4 variables in $\mathcal{M}^*$. Let φ be a first order statement true in $\mathcal{M}^*$. Let $L \supseteq \Psi$ be a finite language contained in the language of $\mathcal{M}^*$ such that φ is a formula of L. We seek a finite structure $\mathcal{M}'$ for the language L such that

1. $\mathcal{M}' \models \varphi$.
2. $\mathcal{M}'$ has at most k 4-types.
3. Every $(Aut\,\mathcal{M}')$-invariant 4-ary relation on $\mathcal{M}'$ is defined by one of the symbols in L.

Note that properties (1–3) taken jointly constitute a standard property of a finite language, and are satisfied (in the internal sense) in a nonstandardly finite structure, hence also in some finite structure. ∎

Lemma 3.5.6. *Let J be a semiprojective or basic linear Lie geometry, $C \subseteq J$ finite, and suppose that $(J;C)$ (C treated as a set of constants) is canonically embedded in the structure $(\mathcal{M};A)$. Let $C' = acl_{\mathcal{M}}(A) \cap J$. Then C' is finite and $(J;C')$ is canonically embedded in $(\mathcal{M};A)$.*

Proof. $C' \subseteq acl(C)$ in the sense of J, so C' is finite.

Let R be an A-definable relation on J. Then R is C-definable and thus $R \in J^{\mathrm{eq}}$. It follows from weak elimination of imaginaries that R is C'-definable. ∎

Lemma 3.5.7. *Let $\mathcal{M}$ be internally finite, J a semiprojective or linear geometry, B-definable, and $C \subseteq J$ finite with (J/C) canonically embedded in $(\mathcal{M}^*;B)$. Assume that C is nondegenerate if J involves a form, and otherwise, if J is pure projective, then assume that in $\mathcal{M}^*$ the definable dual of the linear model V is trivial. Then the group G induced on J by the internal automorphism group of $\mathcal{M}$ over B contains $(Aut(J;C))^{(\infty)}$.*

Proof. In the nondegenerate case, dealing with J over C is equivalent to dealing with J/C and Lemma 3.5.2 of §3.4 applies. In the pure projective case $(Aut(J;C))^{(\infty)}$ has the form $Hom(W,U) \rtimes SL(W)$ relative to a decomposition of the linear model $V = W \oplus U$ with U covering C, and all we learn from looking at the localization is that G induces at least $SL(W)$ on the localization; thus the subgroup of $Hom(W,U) \rtimes SL(W)$ induced by G is $H \rtimes SL(W)$ with H an $SL(W)$-invariant subgroup of $Hom(W,U)$. Then H will be $Hom(W,U_0)$ for some subspace U_0 of U and $P(W \oplus U_0)$ is the unique minimal $G^{(\infty)}$-invariant subspace of J. Thus this space is G-invariant. But as we are in the pure projective case there can be no definable subspace of finite codimension, so $U_0 = U$ and $H = Hom(W,U) \rtimes SL(W)$. ∎

Lemma 3.5.8. *Let $\mathcal{M}$ be an internally finite locally Lie coordinatized structure with respect to the coordinate systems in $\mathcal{J}$ and suppose that*

1. *Whenever $J_b \in \mathcal{J}$ is pure projective, with linear model V, the definable dual V^* is (0).*
2. *Whenever $J_b \in \mathcal{J}$ is symplectic of characteristic 2, there are no definable quadratic forms on J_b compatible with the given symplectic form.*

Then for any finite subset A of $\mathcal{M}$ closed downwards with respect to the coordinatizing tree, we have

3. *For $b \in A$, if J_b is nonaffine, then for some finite subset $C \subseteq J_b$, the structure $(J; C)$ is canonically embedded in $\mathcal{M}^*$ over A.*
4. *For $J_1, J_2 \in \mathcal{J}$ nonaffine, with defining parameters in A, if $C_i = acl_{\mathcal{M}^*}(A) \cap J_i$, then either $(J_1; C_1)$ and $(J_2; C_2)$ are orthogonal over A, or else there is an A-definable bijection of J_1/C_1 with J_2/C_2.*

Proof. We prove (3, 4) simultaneously by induction on the size of A.

Let A, b be given. We prove (3). If A is the branch below b then (3) holds by definition of local lie coordinatization. So we may suppose that A contains elements not on the branch below b; let $c \in A$ be maximal among such elements, and $B = A - \{c\}$. Induction applies to B. In particular $(J_b; C_0)$ is canonically embedded in $\mathcal{M}^*$ over B, for some finite $C_0 \subseteq J_b$. We may take C_0 nondegenerate when a form is present. Then the internal automorphism group of $\mathcal{M}^*$ over B induces at least $(Aut(J_b; C_0))^{(\infty)}$ on J_b.

If c is algebraic over B, then its stabilizer in the internal automorphism group of $(\mathcal{M}^*; B)$ has finite index, hence also covers $(Aut(J_b; C_0))^{(\infty)}$. Thus in this case $(J_b; C_0)$ is canonically embedded in $\mathcal{M}^*$ over A.

Suppose therefore that c is not algebraic over B. Thus there is a geometry J_2 associated to a parameter d of B, with $c \in J_2$. We will write J_1 for J_b. Let $C_i = acl_{\mathcal{M}^*}(B) \cap J_i$. Then $(J_i; C_i)$ is canonically embedded in $\mathcal{M}^*$ over B by Lemma 3.5.6, and (4) applies to this pair if J_2 is also nonaffine.

Case 1. J_2 is nonaffine, and $(J_2; C_2)$ is orthogonal to $(J_1; C_1)$.

Then $(J_1, J_2; C_1 C_2)$ is canonically embedded in $\mathcal{M}^*$ over B and hence $(J_1; C_1)$ is canonically embedded in $\mathcal{M}^*$ over A.

Case 2. J_2 is affine, with corresponding linear geometry V_2, and the projectivization $P_2 = P(V_2/B)$ is orthogonal to J_1/B over B.

As the orthogonality statement is preserved by adding parameters from J_1, and this does not affect the desired conclusion (3), we may take C_1 to be nondegenerate, or J_1 to be pure projective. We now work with the internal automorphism groups.

Let G be the automorphism group of $(J_1; B)$, H the automorphism group of J_2, and $G(X)$ and $H(X)$ the pointwise stabilizers. Then $G(P_2) = G$ since the geometries are orthogonal and basic. Thus we have

$$G/G(J_2) \simeq H(P_2, B)/H(P_2, J_1, B).$$

On the right hand side we have a solvable group and hence $G(J_2)$ contains $G^{(\infty)}$. Thus $(J_1; B)$ is canonically embedded in $(J_1; BJ_2)$ and in particular is canonically embedded over $B \cup \{c\} = A$.

Case 3. J_2 is nonaffine and is nonorthogonal to J_1 over B.

Find $J' = J_{b'}$ with $b' \leq b$ minimal such that J' and J_1 are nonorthogonal. By the induction hypothesis (4) applied to the branch below b, there s a b-definable bijection between J'/b and J_1, which must be an isomorphism of weak geometries. Accordingly, we may replace J_1 by J', and if $b' < b$ conclude by induction. Thus we now assume J_1 is orthogonal to every earlier geometry. In much the same way we may assume that J_2 is orthogonal to every earlier geometry.

As these geometries are nonorthogonal, they are now assumed orthogonal to every geometry associated with a parameter below b or d. It follows that $acl(bd) \cap J_i = \emptyset$ for $i = 1, 2$. The induction hypothesis (4) applies to the union of the branches up to b and d, and gives a bd-definable bijection between J_1 and J_2. Thus $c \in dcl(Bc')$ for some $c' \in J_1$, and (3) follows.

Case 4. J_2 is affine, with corresponding linear geometry V_2; and the projectivization $P(V_2/B)$ is nonorthogonal to J_1 over B.

We minimize parameters as in the previous case, taking J_1 orthogonal to its predecessors, and taking P_2 to be a (nonaffine) geometry nonorthogonal to $P(V_2/B)$ and minimal below d. Then P_2 and J_1 can be identified, as in the previous case, and we apply Lemma 3.5.8 to J_2 and P_2/B. There are then three possibilities.

If J_2 has a 0-definable point in $\mathcal{M}^*$, then $dcl(A) = dcl(B, c')$ for some $c' \in V_2$ and we may replace c by c' and return to the previous case.

If in $\mathcal{M}^*$ we have a B-definable bijection of J_2 with Q, then by hypothesis (2) Q is also part of V_2, and again we reduce to the previous case.

Suppose finally that J_2 is canonically embedded in $\mathcal{M}^*$. Now P_2/B is geometrically definable over J_2, so $(P_2/B, J_2)$ is canonically embedded in $\mathcal{M}^*$. Furthermore, P_2/B is canonically embedded in $(P_2/B, J_2; c)$ (one affine parameter). Thus P_2/B is embedded in $(\mathcal{M}^*; c)$. As P_2 and J_1 are B-definably identified, we wish to show that P_2 is itself canonically embedded in $(\mathcal{M}^*; c)$. When P_2 carries a form then P_2 is geometrically definable from P_2/B and additional parameters from P_2. When P_2 is pure projective it follows from Lemma 3.5.7 that it is canonically embedded in $\mathcal{M}^*$.

This exhausts the cases and proves (3). We now consider (4): so we have J_1, J_2 nonaffine, with defining parameters in A, and $C_i = acl_{\mathcal{M}^*}(A) \cap J_i$.

We apply Lemma 3.5.8 of §3.4. By hypothesis (1) if the geometries involved are pure projective, the polar case cannot arise between them. So either we have an A-definable bijection of J_1/C_1 with J_2/C_2, or these localizations are orthogonal over A.

Suppose therefore that J_1/C_1 and J_2/C_2 are orthogonal over A. Our claim is that then $(J_1; C_1)$ and $(J_2; C_2)$ are orthogonal over A. If J_1 is pure projective then Lemma 3.5.7 applies to give the orthogonality of $(J_1; C_1)$ and J_2/C_2. If J_1 involves a form then consider $G = Aut(J_1; C_1)$ and the pointwise stabilizer $G(J_1/C_1)$. The quotient $G/G(J_1/C_1)$ is solvable and as in Case 2 above it follows that $(J_1; C_1)$ and J_2/C_2 are orthogonal over A. In this case they remain orthogonal over a nondegenerate extension C_1' of C_1 and $(J_1; C_1')$ is definably equivalent to J_1/C_1'.

If J_2 is pure projective the same argument gives us that $(J_1; C_1)$ or $(J_1; C_1')$, as the case may be, is orthogonal to $(J_2; C_2)$. Otherwise, we may suppose that both J_1 and J_2 involve forms, and that $(J_1; C_1')$ is definably equivalent to J_1/C_1, so that repetition of the first argument gives the orthogonality of $(J_1; C_1')$ and $(J_2; C_2)$, using the solvability of the relative automorphism group for $(J_2; C_2)$ over J_2/C_2. By Lemma 3.5.6 the orthogonality holds over A. ∎

Lemma 3.5.9. Let $\mathcal{M}$ be an internally finite locally Lie coordinatized structure. Then $\mathcal{M}^*$ is Lie coordinatizable. If in addition $\mathcal{M}$ is strongly 4-quasifinite then $\mathcal{M}$ is Lie coordinatizable.

Proof. We will apply the previous lemma. The first point is that without loss of generality we may suppose that the coordinatizing family $\mathcal{J}$ satisfies the following:

(i) whenever $J_b \in \mathcal{J}$ is pure projective, with linear model V, the definable dual J^* is (0);

(ii) whenever $J_b \in \mathcal{J}$ is symplectic of characteristic 2, there are no definable quadratic forms on J_b compatible with the given symplectic form.

In other words, if the definable dual J^* is nontrivial, then J is part of a polar geometry encoded in $\mathcal{M}$ which may be used in place of J, and if a symplectic space carries a nontrivial form (and is acted on by the full symplectic group) then it may be replaced by the corresponding quadratic geometry, interpreted in $\mathcal{M}$.

So we have, in particular, the following conclusion from Lemma 3.5.8 for any finite subset A of $\mathcal{M}$:

For $b \in A$, if J_b is nonaffine then for some finite subset $C \subseteq J_b$, the structure $(J; C)$ is canonically embedded in $\mathcal{M}^*$ over A]

Varying A, this implies that the nonaffine geometries are stably embedded in $\mathcal{M}^*$. By Lemma 3.5.8 of §3.4 the same is true for the affine geometries. Thus after replacing the semiprojective geometries with projective ones, $\mathcal{M}^*$ is Lie coordinatized.

If in addition $\mathcal{M}$ is strongly 4-quasifinite, then the Lie coordinatization can be defined using formulas in the language of $\mathcal{M}$. ∎

There has been a certain amount of vacillation between projective and semi-projective geometries visible. The orthogonality theory is simpler for projectives, and elimination of imaginaries holds for the semiprojectives. Furthermore, they are bi-interpretable, so in a sense both theories are available for either version.

We recall the statements of Theorems 2 and 3 of §1.2.

Theorem 3.5.10 (Theorem 2: Characterizations)
The following conditions on a model $\mathcal{M}$ are equivalent:

1. *$\mathcal{M}$ is smoothly approximable.*
2. *$\mathcal{M}$ is weakly approximable.*
3. *$\mathcal{M}$ is strongly quasifinite.*
4. *$\mathcal{M}$ is strongly 4-quasifinite.*
5. *$\mathcal{M}$ is Lie coordinatizable.*
6. *The theory of $\mathcal{M}$ has a model $\mathcal{M}^*$ in a nonstandard universe whose size is an infinite nonstandard integer, and for which the number of internal n-types $s_n^*(\mathcal{M}^*)$ satisfies:*

$$s_n^*(\mathcal{M}^*) \le c^{n^2}$$

   for some finite c, and in which internal n-types and n-types coincide. (Here n varies over standard natural numbers.)

Theorem 3.5.11 (Theorem 3: Reducts). *The following conditions on a model $\mathcal{M}$ are equivalent:*

1. *$\mathcal{M}$ has a smoothly approximable expansion.*
2. *$\mathcal{M}$ has a weakly approximable expansion.*
3. *$\mathcal{M}$ is quasifinite.*
4. *$\mathcal{M}$ is 4-quasifinite.*
5. *$\mathcal{M}$ is weakly Lie coordinatizable.*
6. *The theory of $\mathcal{M}$ has a model $\mathcal{M}^*$ in a nonstandard universe whose size is an infinite nonstandard integer, and for which the number of internal n-types $s_n^*(\mathcal{M}^*)$ satisfies*

$$s_n^*(\mathcal{M}^*) \le c^{n^2}$$

   for some finite c. (Here n varies over standard natural numbers.)

We remarked in §2.1 that weak approximability implies strong quasifiniteness; thus the implications $1 \implies 2 \implies 3 \implies 4$ in Theorem 2 all hold. Furthermore, by existence, finiteness, and homogeneity of envelopes, Lie coordinatizability gives smooth approximation. In the present section we showed that 4-quasifinite structures are Lie coordinatizable. Thus the equivalence of the first five conditions in Theorem 2 has been verified; the estimate needed for the sixth clause will be found in §5.2. One can also verify the equivalence of the first five conditions in Theorem 3 if one replaces *"weakly Lie coordinatizable"* by *"reduct of a Lie coordinatizable structure."* However, the proof that these two conditions are equivalent is subtle and is the subject of Chapter 7.

4

Finiteness Theorems

4.1 GEOMETRICAL FINITENESS

As Ahlbrandt and Ziegler showed, the key combinatorial property of coordinatizing geometries depends on Higman's lemma, itself a special case of the Kruskal tree lemma. This was given an additional degree of flexibility in [HrTC], adequate to our present purposes, once we verify that the geometries we are using possess the following property. The proof is very much the same as in the pure linear case.

Definition 4.1.1. *A countable structure $\mathcal{M}$ is geometrically finite with respect to an ordering $<$ of type ω, if for each n the following holds:*

> *For any sequence of n-tuples $\mathbf{a}_i$ $(i = 1, 2, \ldots)$ in $\mathcal{M}$ there is an order-preserving elementary embedding $\alpha : \mathcal{M} \to \mathcal{M}$ taking $\mathbf{a}_i$ to $\mathbf{a}_j$ for some $i < j$.*

Lemma 4.1.2. *Suppose that $\mathcal{M}$ is $\aleph_0$-categorical and geometrically finite with respect to the ordering $<$. Let $\mathbf{a}$ be a finite sequence of elements of $\mathcal{M}$, and suppose that for each $i = 1, 2, \ldots$ there are given k finite initial segments $S_{i1}, \ldots, S_{ik}$ of $(\mathcal{M}; <)$. Then there is an automorphism α of $\mathcal{M}$, a finite initial segment S of $\mathcal{M}$, and a pair $i < i'$ such that*

1. $\mathbf{a} \subseteq S$; $S_{ij} \subseteq S$ for $j = 1, \ldots, k$.
2. $\alpha^- S$ *is order preserving.*
3. α *fixes* $\mathbf{a}$.
4. $\alpha[S_{ij}] \subseteq S_{i'j}$ *for* $j = 1, \ldots, k$.
5. $\alpha(\max S_{ij}) = \max S_{i'j}$ *for* $j = 1, \ldots, k$.

Proof. Set $b_{ij} = \max S_{ij}$ for all i, j and apply geometrical finiteness to the sequences $(\mathbf{a}, \mathbf{b}_i)$ with $\mathbf{b}_i = (b_{i1}, \ldots, b_{ik})$. The result is an order-preserving elementary map $\beta : \mathcal{M} \to \mathcal{M}$ fixing $\mathbf{a}$ and carrying some $\mathbf{b}_i$ to $\mathbf{b}_{i'}$ with $i < i'$. Restrict β to a large initial segment S of $\mathcal{M}$, and then extend the restriction to an automorphism of $\mathcal{M}$. ∎

In proving the geometrical finiteness of a geometry we first deal with linear models. We work with the following orderings.

Definition 4.1.3. *The* standard *orderings of basic linear (or degenerate) geometries are defined as follows.*

1. *Any ordering of a pure set in order type ω is standard.*
2. *If X is an ordered basis for a vector space V and $<_K$ is an ordering on the base field, with 0 as the first element, then the induced ordering on V is derived from the reverse lexicographic ordering on words in the alphabet K as follows. To any vector v we assign the word consisting of the sequence of its coordinates, truncated after the last nonzero coordinate. A standard ordering of the pure vector space V is any ordering induced by such a pair $(X, <_K)$, where the order type of X is ω.*
3. *If V is a vector space carrying a nondegenerate symplectic or hermitian form, or a nondegenerate quadratic form Q with an associated symmetric form, then an ordered basis X for V will be considered standard if it has the form $(e_1, f_1, e_2, f_2, \ldots)$ where in all cases $(e_i, e_i) = (f_i, f_i) = 0$, $(e_i, f_i) = 1$, the subspaces $H_i = (e_i, f_i)$ are pairwise orthogonal, and in the presence of a quadratic form Q we require furthermore that $Q(e_i) = Q(f_i) = 0$.*

   In this case an ordering on V is considered standard if it is induced by a pair $(X, <_K)$ where X is a standard ordered basis.
4. *A standard ordering of the linear polar geometry (V, W) is defined as in the previous case, using the appropriate version of a standard basis for $V \cup W$; here the e_i form a basis for V, and the f_i form a basis for W.*

We remark that given any standard ordering on a vector space derived from an ordered basis X, the subspaces generated by initial segments of X will be initial segments of V with respect to the induced ordering. We note also that we include the polar case here because it does not reduce to the pure projective case, but we exclude the quadratic case for notational convenience since its underlying set is not a vector space; however, this is a triviality, since after fixing a point of the quadratic geometry it can be treated as an orthogonal geometry.

We review the combinatorial lemma on which geometrical finiteness depends.

Definition 4.1.4. *Let Σ be a finite set.*

1. *A* word *in the alphabet Σ is a finite sequence of elements of Σ. $\Sigma^* = \bigcup_{n \geq 0} \Sigma^n$ is the set of all words in the alphabet Σ.*
2. *The* embeddability ordering *on Σ^* is the partial ordering defined as follows: $w \leq w'$ if there is an order-preserving embedding of w into w'.*
3. *A partially ordered set $(X, <)$ is* well quasi-ordered *if it has no decreasing sequences and no infinite antichains; by Ramsey's theorem, an equivalent condition is that any infinite sequence of distinct elements of X contains an infinite strictly increasing subsequence.*

Fact 4.1.5 (Higman's Lemma [Hi]). *If Σ is a finite alphabet, then the partially ordered set $(\Sigma, <)$, under the embeddability ordering, is well quasi-ordered. Thus for any infinite sequence of words $w^{(i)} \in \Sigma^*$, there is a pair i, j with $i < j$ such that $w^{(i)}$ embeds in $w^{(j)}$.*

We note that this fact is proved more generally in a relative form, for words in any alphabet which is well quasi-ordered, with an appropriately modified embeddability relation. Only the finite case is used here.

Lemma 4.1.6. *The countably infinite versions of the linear and degenerate geometries—a pure set, a pure vector space, a symplectic, hermitian, or orthogonal space, or a polar pair—are geometrically finite with respect to their standard orderings.*

Proof. It will suffice to treat the cases of nondegenerate symplectic, hermitian, or orthogonal spaces, where the notation is uniform. The other nondegenerate cases are simple variations.

We fix a standard ordering $<$ on V with respect to a standard basis $X = (e_1, f_1, \ldots)$ for V and an ordering of K with 0 as initial element. Let $H_i = \langle e_i, f_i \rangle$; this is a nondegenerate plane of the same type as V.

With n fixed we consider n-tuples $\mathbf{a}^{(i)} = (v_{i1}, \ldots, v_{in})$ from V. For each i, expanding relative to the basis X, think of $\mathbf{a}^{(i)}$ as a matrix with n semi-infinite rows, and entries in K. Let $\mathbf{b}^{(i)} = (w_{i1}, \ldots, w_{im_i})$ be the corresponding matrix in reduced row echelon form, and let M_i be the $n \times m_i$ matrix over K connecting the two forms by: $\mathbf{a}^{(i)} = M_i \mathbf{b}^{(i)}$. Without loss of generality, the numbers $m_i = m$ and the matrices $M_i = M$ are independent of i, and we may also suppose that the maps $\mathbf{b}^{(i)} \to \mathbf{b}^{(i')}$ defined by $w_{ij} \mapsto w_{i'j}$ are all isometries with respect to whatever forms are present.

Now we will make the reduction to Higman's lemma, encoding the sequences $\mathbf{b}^{(i)}$ by a word in an appropriate alphabet. We expand each vector w_{ij} as $\sum_r h_{ijr}$ where $h_{ijr} \in H_r$. As the H_r are all isometric we will identify them all with a single plane $H = \langle e, f \rangle$ and consider h_{ijr} to be an element of H. We say that r is the *leading index* for w_{ij} if r is maximal such that $h_{ijr} \neq 0$; we say that the leading index r for w_{ij} is of *type e* if $h_{ijr} \in \langle e \rangle$, and of *type f* otherwise. We associate to $\mathbf{b}^{(i)}$ a sequence $w^{(i)} = (h_{i1}, h_{i2}, \ldots, h_{ir})$ with r the maximal leading index of the w_{ij} in such a way that h_{is} encodes the following sequence of data for $1 \leq j \leq m$:

> the value of $h_{ijs} \in H$;
> whether s is the leading index of w_{ij} (yes/no).

Clearly this information can be expressed by a finite alphabet.

By Higman's lemma we have a pair $i < i'$ such that $w^{(i)}$ embeds in $w^{(i')}$. We will now write out exactly what this means. Let l, l' be the lengths of $w^{(i)}$ and $w^{(i')}$, respectively. Since $i < i'$, there is an increasing function ι :

$\{1, \ldots, l\} \to \{1, \ldots, l'\}$ such that

(1) $$h_{i'\iota(s)} = h_{is} \text{ for } s \le l.$$

or more explicitly, in terms of the data encoded, for $s \le l$ we have

(1.1) $$h_{i'j\iota(s)} = h_{ijs} \text{ for } j \le m.$$

(1.2)   If s is the leading index for w_{ij}, then
$\iota(s)$ is the leading index for $w_{i'j}$.

Set $y_j = \sum\{h_{i'js} : s \notin im\,\iota\}$. The leading index of y_j is less than the leading index of $w_{i'j}$, by (1.2).

     We now associate with ι a linear map β, which is defined on the span of $e_1, f_1, \ldots, e_l, f_l$, as follows:

(2.1) $\beta(e_s) = e_{\iota(s)}$ unless
           s is the leading index of some w_{ij} and is of type e for it.

(2.2) $\beta(f_s) = f_{\iota(s)}$ unless
           s is the leading index of some w_{ij} and is of type f for it.

(2.3) $\beta(h_{ijs}) = h_{i'j\iota(s)} + y_j$ if
           s is the leading index of w_{ij}.

By the initial reduction to row echelon form, a given index s can occur at most once as the leading index of a given type (e or f) for one of the w_{ij}. If s is the leading index for w_{ij} and is of type e for it, then (2.3) and linearity determine $\beta(e_s)$, while if, on the other hand, s has type f for w_{ij}, then (2.3), linearity, and the value of $\beta(e_s)$ determine $\beta(f_s)$. So (2.1–2.3) determine some linear function β. For any r let $H'_r = \bigoplus\{H_s : s < \iota(r), s \notin im\,\iota\}$. Then β has the following properties:

(3.1) $$\beta(h_{ijr}) \in h_{i'j\iota(r)} + H'_r$$

(3.2) $$\beta(w_{ij}) = w_{i'j}$$

     From (3.1) it follows that β is order preserving: if u_1, u_2 have their last difference in the rth component, then $\beta(u_1)$ and $\beta(u_2)$ will differ last in their $\iota(r)$th component, and in the *same manner*. By (3.2) and the relations $\mathbf{a}^{(i)} = M_i \mathbf{b}^{(i)}$, we find $\beta(v_{ij}) = v_{i'j}$.

     It remains to check that β is an isometry. We make use of a basis $X_1 \cup X_2$ for $\langle e_1, f_1, \ldots, e_l, f_l \rangle$ of the following form: X_1 consists of all w_{ij} for $j \le m$; X_2 consists of the e_r and the f_r for which r is not a leading index of corresponding type for any of the w_{ij}. Then by (3.2) β is an isometry on $\langle X_1 \rangle$, and β is also an isometry on $\langle X_2 \rangle$. So we need only check that β preserves inner products between X_1 and X_2 (even in the orthogonal case, this now suffices). In view of the orthogonality of the spaces H_s, the relation (3.1), and the definition of β, this follows.   ∎

Corollary 4.1.7. *The basic geometries are geometrically finite.*

Proof. Let J be the geometry, and V the corresponding linear model, equipped with a standard ordering.

If J is projective order it as follows: $a < b$ if the first representative u for a in V precedes the first representative v of b in V.

If J is affine, then call one element 0, place it first, and order the remainder of J as in V. Similarly, if J is of quadratic type pick one element q of the space Q of quadratic forms on V compatible with the symplectic structure, place it first, and then identify (V, Q) with the orthogonal space $(V; q)$; order it as two copies of a standard orthogonal space. ∎

4.2 SECTIONS

We will establish the notation used in proving that Lie coordinatized structures have finite languages and quasifinite axiomatizations. A particular coordinatization is fixed throughout. The coordinatizing tree, together with some relevant data, will be called the skeleton of the model.

It will be convenient to coordinatize using semiprojectives in place of projectives from this point on.

Definition 4.2.1

1. *A skeletal type consists of the following data:*

a parameter h (the height of a tree);
an assignment τ associating to each i with $1 \leq i \leq h$ the type of a basic semiprojective or affine-with-dual Lie geometry, or a finite structure;
a partial function σ from $\{1, \ldots, h\}$ to $\{1, \ldots, h\}$. Here σ satisfies the following conditions:

   (i) *the domain and range of σ are disjoint and their union is contained in the set of indices i for which $\tau(i)$ is not a finite structure;*

   (ii) *$\sigma(i) < i$;*

   (iii) *the domain of σ contains the set of indices i for which $\tau(i)$ is a basic affine-with-dual Lie geometry.*

A level i for which $\tau(i)$ is a semiprojective type geometry and i is not in the domain of σ is said to be a level of new type.

2. *The skeletal language L_{sk} and skeletal theory T_{sk} associated with a given skeletal type (not shown in the notation) are defined as follows.*

   L_{sk} contains symbols $\leq$ and P_i ($0 \leq i \leq h$) which are asserted by T_{sk} to constitute a tree ordering of height h with levels given by the unary predicates $P_0, \ldots, P_h$; P_0 consists of the root alone. There should also be

predecessor functions for the tree order, so that a substructure will be a subtree.

L_{sk} contains several additional symbols. In the first place, it contains languages suitable to the description of structures of the types specified by the τ-component of the skeletal type. T_{sk} asserts, using these symbols, that the tree successors of a given point at level $i - 1$ form a structure of the type specified by $\tau(i)$, that is, either a specific finite structure or an infinite dimensional basic geometry of specified type. It will be convenient to write $P_i(a)$ for the successors of a point a at level $i - 1$; so T_{sk} controls the type of each $P_i(a)$.

Finally, and crucially, the σ-component of the skeletal type furnishes nonorthogonality information. L_{sk} contains function symbols in several variables f_{ij} whenever $j = \sigma(i)$ representing a parametrized family of functions f_{ija}, where a varies over the points at level $i - 1$, providing a bijection between the projectivization of $P_i(a)$ and a localization of the projectivization of $P_j(a')$ relative to some finite subset, where a' is the element lying below a at height $j - 1$.

It is not quite necessary to fix the skeletal data, as long as the various variables involved, such as the sizes of the finite structures, are kept bounded. However, we can analyze more general situations of this type by dealing with each possible refinement to full skeletal data.

Definition 4.2.2. *Let the skeletal data (h, τ, σ) be fixed, hence also the skeletal language L_{sk} and the skeletal theory T_{sk}. Let L be an expansion of L_{sk}.*

1. *A skeleton with given skeletal data is a model for T_{sk}.*
2. *A skeletal expansion is a structure for the language L whose reduct to L_{sk} is a model of T_{sk}. It has* true dimensions *if not only the type of the geometry, but its isomorphism type, is determined by the atomic type of its controlling parameter.*
3. *A fully* proper *model for the language L is a skeletal expansion which satisfies*

   (i) *The L_{sk}-reduct of each layer $P_i(a)$ with i in the range of σ (that is, the pure geometry) is fully embedded in $\mathcal{M}$.*

   (ii) *If $a' \leq a$ in the tree lie at level $i - 1$ and $j - 1$ respectively, with i, j in the range of σ, then $P_i(a')$ and $P_j(a)$ are orthogonal*

   (iii) *The dual affine part of an affine-with-dual geometry is the full definable affine dual.*

Lemma 4.2.3. *The class of fully proper L-structures relative to a given skeletal theory is an elementary class.*

Proof. The point that requires care is the axiomatization of stable embeddedness of a given geometry J in $\mathcal{M}$, since in order to state in first-order terms the definability of the relativization of a formula φ to J using parameters of J, it is necessary to give an a priori bound on the number of parameters needed in J.

So let $D_b = \{x \in J : \varphi(x, b)\}$ be an $\mathcal{M}$-definable subset of J with parameters b (containing defining parameters for J). If this is J-definable, it is definable using parameters in $J \cap acl(b)$, by weak elimination of imaginaries. This is a finite dimensional subspace of J of dimension at most $rk(b)$, and $rk(b)$ is at most the height h times the number of entries in the sequence b. ∎

We now deal at length with skeletons and expansions of skeletons. View L_{sk} and T_{sk} as fixed for the present.

Definition 4.2.4. *Let $\mathcal{M}$ be a countable skeletal expansion.*
An Ahlbrandt–Ziegler enumeration (or more specifically, a breadth-first Ahlbrandt–Ziegler enumeration) is an enumeration of $\mathcal{M}$ derived from some data of the following type, according to the recipe following. The data will be

1. *A standard enumeration of the projectivization of each one of the semiprojective layers at level i where i is a level of new type;*
2. *An enumeration of each of the finite structures found in the coordinate tree;*
3. *A set $C_i(a)$ of at most $|K|$ elements (K is the base field) in each of the components $P_i(a)$ of the ith layer, whenever $P_i(a)$ is not finite, chosen so that*

>   if $P_i(a)$ is semiprojective, then $C_i(a)$ is the set of semiprojective points above some point of the projectivization of $P_i(a)$ (in the sense explained below); if $P_i(a)$ is affine then $C_i(a)$ enumerates an affine line in $P_i(a)$.

Relative to these data, we order $\mathcal{M}$ as follows. Enumerate successive layers of the tree; the order in which the ith layer is enumerated is determined first by the enumeration of the previous layer, and for a fixed element a of layer $i - 1$, either

- *the enumeration of $P_i(a)$ is given as part of the data, using one of the clauses $(1, 2)$, or*
- *in the event that $j = \sigma(i)$ is defined, the enumeration of $P_i(a)$ is determined by the enumeration of $P_j(a')$ where a' lies below a at level $j - 1$, as follows. We have by hypothesis a specific identification of the projectivization P_a of $P_i(a)$ with a localization $P_{a'}$ of $P_j(a')$. If $P_i(a)$ is semiprojective then enumerate the points of $C_i(a)$ first; then over these points there is a definable function from the projectivization onto $P_i(a)$, so an ordering of the rest of $P_i(a)$ is determined by an ordering on the corresponding localization of $P_j(a')$ where $j = \sigma(i)$ and a' lies below*

a at level $j - 1$. Such an ordering on the localization of $P_j(a')$ can be induced from the ordering of $P_j(a')$ using first representatives, as in the original discussion of geometrical finiteness. If $P_i(a)$ is affine-with-dual then the dual part is enumerated first, following the enumeration of the projective dual (which is part of the corresponding projective geometry), and then the affine part is enumerated by taking the affine line $C_i(a)$ of (3) first, after which one follows the enumeration of its projectivization as in the semiprojective case.

Definition 4.2.5. *Let M be a countable skeletal expansion.*

A section *of M is an initial segment of M with respect to an Ahlbrandt–Ziegler enumeration. The* height h *of a section is the least level not completely contained in the section. According to this definition the height of M itself should be considered to be undefined.*

Definition 4.2.6. *Let M be a countable skeletal expansion and U a section of M of height h.*

A support *for U consists of the following data (B, a, C) :*

1. *The sequence $B = (B_1, \ldots, B_h)$, with B_i consisting of all points a at level i for which a lies below some point of U at level h, and the tree predecessor of a lies below some point at level h not in U;*
2. *If $i \leq h$ is maximal such that B_i is nonempty: let $a = (a_0, a_1, \ldots, a_{i-1})$ be the (unique) branch leading to B_i;*
3. *If $P_i(a_{i-1})$ is finite let $C_i(a)$ be the complete enumeration of $P_i(a_{i-1})$; if $P_i(a)$ is semiprojective or if B_i meets the affine part, let $C_i(a)$ be the finite subset chosen originally in the construction of the order from which U was derived; if $P_i(a)$ is an affine-with-dual pair and B_i is contained in the affine dual, let $C_i(a)$ be an enumeration of the points of B_i which lie over the last point of the projectivization (the point being that the ordering of the projectivization does not define a unique ordering of the affine dual, but knowing $C_i(a)$ and the projective ordering, the initial segment of the affine dual is determined).*

Note here that a section does not quite determine its support, since the same section may be derivable from different orderings; this is just an abuse of language, and in any case in practice supports are used to determine sections, rather than the reverse.

Lemma 4.2.7. *Let (B, a, C) be given with $B = (B_1, \ldots, B_h)$ a sequence of subsets of the first $h + 1$ layers of a countable skeletal expansion M, $a = (a_0, a_1, \ldots, a_{h'-1})$ the branch leading to $B_{h'}$, where h' is maximal such that this is nonempty, and $C = (C_1, \ldots, C_{h'})$ a sequence of finite enumerated subsets C_i of $P_i(a_{i-1})$. Then whether (B, a, C) is a section support or not is determined by its type in L_{sk}, and if this is so, then the* *section U supported by it consists of everything of level less than h together with everything of level h that lies above an element of one of the sets B_i.*

Furthermore, a and C are of bounded size, and allow B to be recovered from data of the form $(B'_i; B_{ij})_{i \text{ newtype}}$, where B'_i is a finite subset of $P_i(a)$ for i of new type and B_{ij} is a sequence of subsets of B'_i.

Proof. The last paragraph is really the key. In the case in which we are in fact dealing with a section support, the B_{ij} should be the initial segments at level i gotten by projecting the B_j when $h(j) = i$ (but in the affine-with-dual case B_j is either a finite subset of the dual part, or the whole dual component plus a finite subset of the affine part, and in the present context one should throw away the affine dual part if it is completely contained in B_j), and B'_i should be their union (i.e., the longest one).

To determine whether we actually have a section support, what we must determine is whether a candidate sequence B_{ij} of finite subsets of a geometry does, in fact, constitute a sequence of initial segments of that geometry with respect to some standard ordering.

An initial segment of a standard ordering on one of the projective geometries contains an initial segment of the standard basis from which the ordering was defined; conversely, if such a finite basis is found in the set B'_i, isomorphic to an initial segment of a standard basis, and making all B_{ij} initial segments in the induced ordering (relative to some ordering of the base field), then it can be completed to a standard basis for the whole space, for which the given sets constitute initial segments. ∎

Definition 4.2.8. *A* reduced section support *is a sequence B of sequences $B_i = (B_{ij})$ for i of new type and $j = i$ or $\sigma(j) = i$, together with auxiliary data (of bounded size)* **a**, *$C_i(a)$ $(a \in$* **a**$)$ *as in the previous lemma, and the maximal elements a_{ij} of the B_{ij} in a standard ordering of B_i. The $C_i(a)$,* **a**, *and a_{ij} will be called the* bounded part *of the section support.*

Remarks 4.2.9
When the standard ordering on the projectivizations of the $P_i(a)$ is fixed, the B_{ij} are determined by B_i and the bounded part, specifically the a_{ij}.

Sections are atomically L_{sk}-definable from their reduced section supports. We may speak also of sections and section supports in envelopes of Lie co-ordinatized structures, as they can be described in terms of their atomic L_{sk} types.

4.3 FINITE LANGUAGE

Definition 4.3.1. *Let $\mathcal{M}$ be a fully proper countable skeletal expansion. Triples (E, X, e) with E an envelope for $\mathcal{M}$, $X \subseteq E$, and e a finite se-*

quence of elements of E, will be partially ordered by the following relation:

$(E, X, e) \leq (E', X', e')$ if and only if there is an elementary
map $f : E \to E'$ for which $f[X] \subseteq X'$ and $f(e) = e'$.

The partial orderings of interest to us here will be restrictions of this
ordering to the sets $\mathcal{U}_n$ and $\mathcal{S}_n$ of triples in which a has length n and X is,
respectively, a section U of E or a reduced section support S for E.

Lemma 4.3.2. *Let $\mathcal{M}$ be a proper countable skeletal expansion. Suppose that*
$(a_0, a_1, \ldots, a_h)$ *is a branch of the tree, and α_i is an automorphism of the*
$P_i(a_{i-1})$ *for i of new type. Then the union of the α_i is an elementary map*
in $\mathcal{M}$.

Proof. Full embedding and orthogonality. The orthogonality theory applies directly to the projectivizations, but the semiprojective geometries are definable
over them and have the same automorphism group. ∎

Lemma 4.3.3. *Let $\mathcal{M}$ be a proper countable skeletal expansion. The partial*
orderings defined above on $\mathcal{U}_n$ and $\mathcal{S}_n$ are well quasi-orderings.

Proof. The result for reduced section supports implies the result for sections,
so we focus on $\mathcal{S}_n$. We can drop the envelope E from the triple, since given
(E, B, a) and (E', B', a') with E a μ-envelope, E' a μ'-envelope, and $\mu(J)$
embedding in $\mu'(J)$ everywhere, and an elementary map f with $f[B] \subseteq B'$
and $f(a) = a'$, there is an elementary map $E \to E'$ extending it, by (essentially) Lemma 3.2.4. We may thin the original sequence so that the condition
on comparability of μ and μ' holds everywhere.

We treat the case of reduced section supports. This is done as in [HrTC,
Lemma 2.10], which, however, makes use of rather abstract notation for part
of the situation.

Increasing n slightly, we may suppose that the bounded part of the reduced
section support is encoded in a. Now take a sequence $S_k = (B^{(k)}, a^{(k)})$ of
reduced section supports with auxiliary data. Adjusting by automorphisms of
the geometries, using the previous lemma, we may suppose that the orderings
used on the projective geometries of new type are fixed standard orderings, so
that the terms B_i (which initially are sequences (B_{ij})) can be thought of as
initial segments of these geometries. Moving up through the levels i which
are of new type, and thinning the sequence S_k at each stage, we will construct
the desired elementary maps in stages. What we require at stage i is that the
maps be defined through the ith level, be order-preserving on each projective
geometry associated with a level of new type, and fix the data in $a^{(k)}$ occurring
up to the ith level. We require of the sequence S_k that the type of $a^{(k)}$ over
$\bigcup_j P_j(b_{j-1}^{(k)})$ (with $b^{(k)}$ the branch being followed by the $B_i^{(k)}$) be fixed. If this
is the case at a given stage, it can be preserved without difficulty up to the next

new level i. At such a new level i, the elementary maps will have to be chosen carefully to preserve the types of $a^{(k)}$ over the union including the ith level.

Let $A_k = \bigcup_{j<i} P_j(b^{(k)}_{j-1})$. The type of $a^{(k)}$ over $A_k \cup P_i(b^{(k)}_{i-1})$ is determined by its (known) type over A_k and its type over $c_k = acl(a^{(k)}) \cap P_i(b^{(k)}_{i-1})$. So we impose on our elementary maps the additional constraint that they preserve the c_k. Exactly this condition is allowed by geometrical finiteness, after thinning the sequence S_k (and applying Ramsey's theorem): for $k < l$ we may carry $B^{(k)}_i$ into $B^{(l)}_i$ by an order-preserving elementary map which carries c_k to c_l. Thinning down so that the types of the $a^{(k)}$ over the c_k correspond, this completes the ith stage. ∎

Lemma 4.3.4. *Let E be an envelope, U a section of E, and E' an envelope contained in E, with the support S of U contained in E'. Then $E' \cap U$ is the section of E' supported by S.*

Proof. The statement is a bit misleading; the issue is not so much whether S supports $E' \cap U$, but rather whether S fulfills the definition of section support relative to E' in the first place. This is essentially one of the points made in Lemma 4.2.7. In the present version, the statement is that if B is an increasing sequence of initial segments of a projective Lie geometry J, with respect to some standard ordering, and lies in a subgeometry J' of J, then B is also a sequence of initial segments of J' with respect to a standard order, the point being that an initial segment of an appropriate standard basis can be extracted from B and completed in J or J'. ∎

Lemma 4.3.5. *Let $\mathcal{M}$ be a Lie coordinatized structure. Then there is an integer k with the following properties:*

1. *For any envelope E, any section U of E, and any $a \in E$, if $a \in acl(U)$ then for some subset C of U of size at most k, a is algebraic over C and its multiplicity over U and over C coincide.*
2. *For any envelope E, any section support S in E, and any $a \in E$, if $a \in acl(S)$ then for some subset C of S of size at most k, a is algebraic over C and its multiplicity over S and over C coincide.*

Proof. The contrary to (1) would yield as a counterexample a sequence (E_k, U_k, a_k) refuting the claim for each k. After passing to a subsequence and applying Lemma 4.3.3 we get a single element a algebraic over an increasing chain of sets U_{k_i} but whose type over U_{k_i} cannot be fixed by k_i elements. The multiplicity m of a over $\bigcup_i U_{k_i}$ is of course the same as its multiplicity over some finite set C contained in all U_{k_i} from some point on, and once $k_i > |C|$ we reach a contradiction.

The failure of (2) is refuted similarly. ∎

Definition 4.3.6. *The* standard language *for a Lie coordinatized structure will be the language L containing all 0-definable $(k+1)$-ary predicates with k (minimal) furnished by the preceding lemma. Note that $k \geq 2$.*

Proposition 4.3.7. *Every Lie coordinatized structure $\mathcal{M}$ admits a finite language L. The standard language will do. The standard language also satisfies the following homogeneity conditions:*

1. *Every section of any envelope of $\mathcal{M}$ is L-homogeneous: if E is an envelope of $\mathcal{M}$, U a section of E, and $f : U \to \mathcal{M}$ an L-map, then f is elementary.*
2. *Every section support of any envelope of $\mathcal{M}$ is L-homogeneous in the same sense.*

Proof. Let L be the standard language for $\mathcal{M}$. Part (1) includes the statement that the language L is adequate for $\mathcal{M}$. We use semiprojectives rather than projectives in the coordinatization.

Both (1) and (2) reduce to finite envelopes, using Lemma 4.3.4. We can enumerate the envelope E so that any initial segment of E is a section. Here we are viewing the envelope as a subset of a coordinatized structure (in the construction of envelopes, we added some sorts of $\mathcal{M}^{\mathrm{eq}}$). Whenever we encounter an affine point the whole dual-affine part is already in the part enumerated. For (1) it suffices to show

(1′) For any section U of an envelope E of $\mathcal{M}$, and a the next element of E, the L-type of a over U determines its type over U.

In the algebraic case this holds by the choice of k. In the nonalgebraic case the L-type of a over U ensures that a is nonalgebraic, again by the choice of k. Let P be the component of the coordinatizing tree in which a lies. We claim that

(∗) $acl(U) \cap P \subseteq U.$

As a is not algebraic over U, P is neither finite nor a semiprojective geometry "repeating" an earlier one. Thus it is either a semiprojective geometry of new type or an affine-with-dual pair. Consider the affine case. Again by the nonalgebraicity assumption, U will contain no affine point of P, while a is affine; as a is the next point of the enumeration, U contains the full dual-affine part of P in E, and as E is itself algebraically closed in $\mathcal{M}$, the claim (∗) holds in this case. Suppose now that P is semiprojective of new type, so orthogonal to all projective geometries J' at lower levels. Then $acl(U) \cap P = acl(U \cap P) \cap P$. This reduces our claim to the corresponding claim (∗) in a single geometry, where it is a property of standard enumerations.

This gives (∗). Now in $\mathcal{M}$ as P is fully embedded, the type of a over $acl(U) \cap P$ implies its type over U, and by (∗) $acl(U) \cap P$ is $U \cap P$. To conclude, then, it suffices to observe that $tp_k(a/U \cap P)$ proves $tp(a/U \cap P)$, which holds since $k \geq 3$ and P is a-definable (directly from the tree language, in fact).

For (2) we may proceed similarly, extending f over an enumeration of E. ∎

Lemma 4.3.8. *Let M be Lie coordinatized, L the standard language for M. Then for any section U of any envelope E, the theory of U is model complete.*

Proof. We must show that any type in U is equivalent to an existential type. We show by induction on the section U:

$(*)$ For any finite sequence c in U there is a finite sequence c' in U such that $tp_L(cc')$ implies $tp_M(c)$.

Granted this, if c is expanded first to contain a support for U, then the type of c in M will determine its type in U, and our claim follows.

This statement passes through at limit stages, so we deal with the case $U = U_1 \cup \{a\}$. We may suppose $c = c_1 a$ with c_1 from U_1. We need first a finite set C such that $tp_L(a/C)$ determines $tp(a/c_1)$. This is a consequence of $(1')$ from the previous proof. (C will grow with c_1 in general, when a is the first affine point.) We may suppose $c_1 \subseteq C$.

At this point, it is useful to make the statement

$$\text{``}tp_L(a/C) \text{ determines } tp(a/c_1)\text{''}$$

more explicit. This is a statement belonging to the type of C; another way of putting it is that the type of C and the L-type of a over C determine the type of $c_1 a$.

We let C' be chosen by applying $(*)$ inductively to C and U'. We claim that $tp_L(CaC')$ determines $tp_M(c_1a)$. Given $tp_L(CaC')$, we first recover $tp_M(C)$. Then we know that $tp_L(aC)$ determines $tp_M(c_1a)$. ∎

4.4 QUASIFINITE AXIOMATIZABILITY

In this section we provide reasonably explicit axiomatizations of theories of Lie coordinatized structures, modulo certain information which is determined only qualitatively by the geometrical finiteness of the coordinatizing geometries.

Definition 4.4.1. *Let M be Lie coordinatized and L a specified language for M. A characteristic sentence for M is an L-sentence whose countable models which are skeletal expansions with true dimensions are exactly the envelopes of M and their isomorphic images.*

Lemma 4.4.2. *Let a skeletal type and corresponding skeletal language L_{sk} be fixed. For any k there is a (uniformly computable) integer k^* such that any $2k$ elements of a section U of a skeleton $\mathcal{M}$ for L_{sk}, with support S, are contained in a subsection U' whose support S' has the same bounded part and satisfies $|S'| \leq k^*$.*

Proof. Note that the subsection will be taken with respect to a different ordering.

This statement reduces to the same statement in a single projective geometry. The existence of k^* follows from the geometric finiteness. Its computability follows from the decidability of the theory of the geometry.   ∎

Proposition 4.4.3. *Let a skeletal type and corresponding skeletal language L_{sk} be fixed, and let L be a finite language containing L_{sk}. Then there is a recursive class Ξ of (potential) characteristic sentences, which can be found uniformly in the data L_{sk}, L, with the following properties:*

1. *If $\mathcal{M}$ is a skeletal expansion with true dimensions relative to L_{sk}, and $\mathcal{M} \models \xi$ (some $\xi \in \Xi$), then every countable model of ξ with true dimensions is isomorphic with an envelope of $\mathcal{M}$.*
2. *Any Lie coordinatized structure with coordinatizing skeleton M_{sk} satisfies one of the sentences in Ξ.*

   In particular, every Lie coordinatized structure has a characteristic sentence.

Proof. We form the set Ξ^* of sextuples $(\xi, k, k^*, k^{**}, L', \Sigma)$ satisfying the following six conditions, and then take Ξ to consist of the sentences ξ for which some suitable k, k^*, k^{**}, L', and Σ can be found; this will make Ξ recursively enumerable but by a standard device any r.e. set of sentences is equivalent to a recursive set: it suffices to replace each sentence ξ by a logically equivalent one whose length is at least the time taken to enumerate ξ.

The conditions on $(\xi, k, k^*, k^{**}, L', \Sigma)$ are as follows:

(i) L' is a list of formulas of L, each with at most $k + 1$ free variables. L' is to be thought of as a new language, and the given formulas will be called L'-atomic formulas. These formulas will include the atomic formulas of L. Σ is a finite set of existential L'-formulas.

(ii) ξ implies the skeletal theory T_{sk}, apart from the clause asserting infinite dimensionality of certain geometries.

(iii) ξ asserts that certain quantifier free L'-formulas in $k + 1$ free variables are algebraic in the last k variables, that is for each choice of these k variables, the formula has only finitely many solutions (with a specified bound). These formulas will be called *explicitly algebraic*.

(iv) For any $\forall\exists$ L'-sentence with k^* universal quantifiers and $k+1$ existential ones, ξ specifies the truth or falsity of the statement.

(v) For any section support S of size $l \leq k^*$ whose atomic L'-type is r (in l variables), and for any L'-formula φ in these l variables with at most $k + 1$ quantifiers, ξ implies that either all realizations of r satisfy φ, or all realizations of r satisfy $\neg\varphi$.

(vi) For any section U of a model $\mathcal{M}$ of ξ with support S of size at most k^*, and any $a \in \mathcal{M}$, ξ asserts that one of the following occurs (to be elucidated more fully below):

    (vi.a) There is a set $B \subseteq U$ of order at most k for which the quantifier-free L'-type of a over B is explicitly algebraic and "implies its L'-type over U";

    (vi.b) a lies in an affine-with-dual geometry J whose dual affine part D (if present) lies in U, and the geometric type of a over D "implies its L'-type over U."

    (vi.c) a lies in a semiprojective geometry of new type J and the geometric type of a over $J \cap U$ "implies its L'-type over U."

It remains to formalize condition (vi) more completely, and in so doing to explain the role of the formulas in Σ. We are dealing with expressions of the form "ξ states that $tp^*(a/X)$ determines $tp(a/U)$" where the second type is an atomic L'-type and the first type is some part of an atomic L'-type.

To formalize (vi.a) we consider a formula $\alpha(x; y)$ expressing the atomic L'-type of a over B, $|B| \leq k$, with x standing for a and y for B, and we consider any other formula $\beta(x; y')$ in $l \leq k$ variables. We are trying to formalize (and to put into ξ) the statement $(\alpha \implies \beta)$, whenever this is true. This is done as follows, elaborating on the model completeness:

    (vi.a') For any $B' \subseteq U$ with $|B'| = l$ (enumerated as a sequence of length l), and any section support $S' \subseteq S$ with $|S'| \leq k^*$ such that the section U' supported by S' contains $B \cup B'$: if $\beta(a, B')$ holds then there is an existential formula $\sigma(z, y, y')$ in Σ where z corresponds to an enumeration of S', true in U', such that ξ implies that $[\sigma(z, y, y') \& \alpha(x, y)] \implies \beta(x, y')$.

The existential quantifiers in σ will refer to the section supported by z. We treat (vi.b) and (vi.c) similarly, e.g.:

    (vi.b') For any $B' \subseteq U$ with $|B'| = l$ (enumerated as a sequence of length l), and any section support $S' \subseteq S$ with $|S'| \leq k^{**}$ such that the section U' supported by S' contains the affine dual of the component of a and B': if $\beta(a, B')$ holds then there is an existential formula $\sigma(z, y, y')$ in Σ where z corresponds to an enumeration of S' and y enumerates some elements of the affine dual part, that σ holds in U' and ξ implies that $[\sigma(z, y, y') \& \alpha(x, y)] \implies \beta(x, y')$.

We require of course that for every β involving k variables there should be a suitable α for which the corresponding version of (vi) holds. This can be viewed as a condition on k^* and k^{**}, particularly when we wish to verify point (2).

We claim that with this choice of ξ, (1,2) hold. We begin by commenting on (2), which amounts to an elaboration of the proof of the existence of a finite language. The parameter k is the one used to define a standard language, and L' is the standard language, given in terms of 0-definable relations in the specified language L. Clause (iii) is natural in view of the definition of k; given $\mathcal{M}$, all the formulas of the given type which are algebraic in $\mathcal{M}$ will be made explicitly algebraic. Point (v) reflects the homogeneity of section supports. Finally, point (vi) reflects the control of types over envelopes, and the model completeness of the theory of the envelopes. Part (vi.a) is an accurate reflection of the role of k as a bound for the base of algebraicity over an envelope. Point (vi.b) requires further elucidation. We will have in general $tp_G(a/D) \vdash tp_{L'}(a/U)$ ("G" for "geometric"). Now $tp_{L'}(a/U)$ consists of formulas β of the appropriate form for (vi.b'). The formulas $\alpha(x, y)$ coming from $tp_G(a/D)$ may require more than k variables. However, given $\mathcal{M}$, there will be a bound k_1 for the number of variables needed, and a corresponding bound k^{**} for the size of a section support needed to capture $k_1 + k$ variables. Then (vi.b') expresses (vi.b).

We turn to (1): $\mathcal{M}$ is a proper L-structure relative to L_{sk} and $\mathcal{M} \models \xi$ (some $\xi \in \Xi$). We claim that every countable model $\mathcal{M}'$ of ξ is isomorphic with an envelope of $\mathcal{M}$ (or with the restriction of an envelope in an adequate expansion of $\mathcal{M}$, to the sorts of $\mathcal{M}$).

If $\mathcal{M}^*$ is an $\aleph_1$-saturated elementary extension of $\mathcal{M}$ then $\mathcal{M}$ is the countable envelope for $\mathcal{M}^*$ with all μ-invariants infinite dimensional. It suffices to show that $\mathcal{M}'$ is isomorphic with an envelope in $\mathcal{M}^*$.

We enumerate $\mathcal{M}'$ so that each initial segment is a section of the skeleton, and we define a map $F : \mathcal{M}' \to \mathcal{M}^*$ by induction. An *approximation* to F will be a pair (f, U) satisfying the following three conditions:

(a) U is a section of $\mathcal{M}'$ with support S;

(b) f is an L'-embedding of U into $\mathcal{M}^*$;

(c) If J_b is a semiprojective component of $\mathcal{M}'$ of new type, with $b \in U$, $J_b \subseteq U$, then $acl(f[U]) \cap J_{f(b)}$ is $f[J_b]$.

Condition (c) essentially rules out "accidents" in which as f is extended, some new value generates a coordinate in a geometry which has already been dealt with. Since we have been rather more careful in the axiomatization to specify what is algebraic than we have been to avoid algebraicity, there is something to be concerned with.

If we are able to carry out the inductive step in which a single element is added to U, then the construction passes smoothly through limit stages and produces a total $(F, \mathcal{M}')$ satisfying the conditions (b, c) with $U = \mathcal{M}'$. By (c) the image of F will be algebraically closed in each semiprojective component of new type coded by an element of the image. It follows easily that $F[\mathcal{M}']$ is algebraically closed in $\mathcal{M}^*$. Also if $c \in \mathcal{M}^* - F[\mathcal{M}']$ then there is c' definable from c with the same property, lying in a semiprojective component of new type, whose defining parameter is in the image of F. Again (c) applies and leads to the maximality clause in the definition of envelope after passing to the canonical projective associated with the given component (one of the sorts which should be added to $\mathcal{M}$ in an adequate expansion).

The last point is that the isomorphism type of a coordinatizing component of $F[\mathcal{M}']$ with a given defining parameter b is constant over all conjugates of b (in $\mathcal{M}^*$) lying in the image. This follows since F is an L-embedding.

So what remains to be checked is the extendability of an approximation (f, U) to the next element a of $\mathcal{M}'$. Let J be the component of $\mathcal{M}'$ in which a lies. Then the L'-type of a is determined either by an explicitly algebraic formula ψ, or a geometric type over part of U. We extend f by letting $f(a)$ be any realization of the corresponding type in $\mathcal{M}^*$. If a is explicitly algebraic then condition (v) implies that $\mathcal{M}^*$, a model of ξ will realize this type. If a is geometric, then $\mathcal{M}^*$, being a Lie coordinatized model in the first place, will realize the appropriate type, using saturation. Let the extension be denoted (f', U'). We claim that the conditions (b, c) are preserved.

Condition (b) is controlled by properties (vi.a, vi.b) of ξ. Note here that the auxiliary formulas in Σ are existential and hence are preserved by embedding.

The condition (c) is obviously preserved if a is algebraic over U or more generally if $acl\, f[U'] \cap J_{fb} = acl\, f[U] \cap J_{fb}$. So we must consider the case in which a is not algebraic over U but some element of J_{fb} not in $acl\, f[U] \cap J_{fb}$ becomes algebraic over $f[Ua]$. Let S be the support of the section U, and let U^* be the section of $\mathcal{M}^*$ supported by $f[S]$, which contains J_{fb} in particular. Then fa is algebraic over U^* and hence is k-algebraic over some section of $\mathcal{M}^*$ whose support $fS' \subseteq fS$ is of size at most k^*. Accordingly ξ asserts some element a' of the geometry J containing a in $\mathcal{M}$ will be algebraic over the section U' supported by S'. In particular $acl(U)$ meets J. On the other hand $a \notin acl(U')$. Thus J is a new geometry and by orthogonality theory in $\mathcal{M}^*$, $acl\, f[Ua] \cap J_{fb} = acl\, f[U] \cap J_{fb}$. ∎

4.5 ZIEGLER'S FINITENESS CONJECTURE

Proposition 4.5.1. *Let a skeletal type and corresponding skeletal language L_{sk} be fixed, and let L be a fixed finite language containing L_{sk}. Then there are only finitely many Lie coordinatized structures in the language L having a given skeleton M_{sk}, up to isomorphism.*

Proof. It suffices to combine Proposition 4.4.3 with the Compactness Theorem. For this one must check that the class of Lie coordinatized structures in the language L with the specified skeleton is an elementary class. Thus one must review the various conditions involved in Lie coordinatization.

Note that the skeleton fixes the language of the individual geometries. In particular, the notion of canonical embedding is first order, as is the notion of orientability.

One must also express the condition of stable embedding for the geometries. We can use Lemma 2.3.3. Thus it suffices to bound the size of $acl(a) \cap J$ uniformly. But $|acl(a) \cap J|$ has dimension at most the height of the skeleton times the length of a.

Thus compactness applies.   ∎

Definition 4.5.2. *Let $\mathcal{M}$ be a structure.*

1. A cover *of $\mathcal{M}$ is a structure $\mathcal{N}$ and a map $\pi : \mathcal{N} \rightarrow \mathcal{M}$ such that the equivalence relation E_π given by "$\pi x = \pi y$" is 0-definable in $\mathcal{N}$, and the set of E_π-invariant 0-definable relations on $\mathcal{N}$ coincides with the set of pullbacks along π of the 0-definable relations in $\mathcal{M}$.*

2. Two covers $\pi_1 : \mathcal{N}_1 \rightarrow \mathcal{M}$, $\pi_2 : \mathcal{N}_2 \rightarrow \mathcal{M}$ are equivalent *if there is a bijection $\iota : \mathcal{N}_1 \leftrightarrow \mathcal{N}_2$ compatible with π_1, π_2 which carries the 0-definable relations of $\mathcal{N}_1$ onto those of $\mathcal{N}_2$.*

3. If $\pi : \mathcal{N} \rightarrow \mathcal{M}$ is a cover, then $Aut(\mathcal{N}/\mathcal{M})$ is the group of automorphisms of $\mathcal{N}$ which act trivially on the quotient $\mathcal{M}$. Thus $Aut(\mathcal{N}/\mathcal{M}) \leq \prod_{a \in M} Aut_\mathcal{N}(C_a)$ where $C_a = \pi^{-1}(a)$ and $Aut_\mathcal{N}(C_a)$ is the permutation group induced by the setwise stabilizer of C_a in $Aut\,\mathcal{N}$.

The problem of the theory of covers is to classify or at least restrict the possible covers with given quotient and specified fiber; that is, typically the structures $(C_a, Aut_\mathcal{N}(C_a))$ are specified in advance and are essentially independent of a. Any automorphism group will be a closed subgroup of the symmetric group (in the topology of pointwise convergence with the discrete topology on the underlying set); by the finiteness of language, in the Lie coordinatized case it is even k-closed for some finite k: any permutation which agrees on every set of k elements with an automorphism is itself an automorphism. in the $\aleph_0$-categorical context, furthermore, $Aut\,\mathcal{N}$ induces $Aut\,\mathcal{M}$; in particular, if the automorphism group of the fibers is abelian, then $Aut(\mathcal{N}/\mathcal{M})$ is an $Aut(\mathcal{M})$-invariant subgroup of the product.

Proposition 4.5.3. *Let $\mathcal{M}$ be a fixed Lie coordinatized structure and let J be a fixed geometry or a finite structure. Then there are only finitely many covers $\pi : \mathcal{N} \rightarrow \mathcal{M}$ up to equivalence which have fiber J and a given relative automorphism group $Aut(\mathcal{N}/\mathcal{M}) \leq \prod_{N/E} Aut\,J$.*

Proof. We apply Proposition 4.5.1. The skeleton $\mathcal{N}_{\mathrm{sk}}$ of $\mathcal{N}$ is determined by the given data and thus it suffices to find a single finite language L adequate for all such covers $\mathcal{N}$. Thus it suffices to bound the arity k of L and the number of k-types occurring in $\mathcal{N}$.

We deal first with the arity, using the language of permutation groups. We must find a bound k so that $Aut(\mathcal{N})$ is a k-closed group, for all suitable covers $\mathcal{N}$. $Aut(\mathcal{M})$ is k_o-closed for some k_o. If we restrict attention to $k \geq k_o$, then $Aut(\mathcal{N})$ is k-closed if and only if $Aut(\mathcal{N}/\mathcal{M})$ is k-closed, as is easily checked. (Note that $Aut\,\mathcal{N}$ induces $Aut\,\mathcal{M}$ by $\aleph_0$-categoricity.)

Thus for $k \geq k_o$ the choice of k is independent of the cover, as long as the relative automorphism group is fixed in advance.

Now with k fixed, consider the number of k-types available in $\mathcal{N}$. If the fiber is finite of order m, then each k-type of $\mathcal{M}$ corresponds to at most m^k k-types of $\mathcal{N}$, so we have the desired bound in this case.

If the fiber is a geometry, to bound the number of k-types we proceed by induction, bounding the number of 1-types over a set A of size j for $j < k$. The 1-type of an element a of the geometry J_b over A is determined by its restriction to the algebraic closure of A in a limited part of J_b^{eq}, e.g. in the affine case the linear version must also be taken. It suffices therefore to bound the dimension of $acl(a) \cap J$ for geometries J associated to J_b. As $rk(Aa/\pi[A]a) \leq j$, the space $acl(\pi[A]a)$ has codimension at most j in $acl(Aa) \cap J$ and thus the desired bound for $\mathcal{N}$ can be given in terms of the data for $\mathcal{M}$. ∎

Remark 4.5.4. In cohomological terms, if $Aut\,J$ is abelian this may be expressed by:
$$H_c^1(Aut\,\mathcal{M}, (\textstyle\prod_{\mathcal{M}} Aut\,J)/K) \text{ is finite}$$

for $K \leq \prod_{\mathcal{M}} Aut\,J$ closed and $(Aut\,\mathcal{M})$-invariant. Cf. [HoPi].

For a more algebraic approach to this type of problem, due to David Evans, see the paper [Ev].

5

Geometric Stability Generalized

5.1 TYPE AMALGAMATION

Definition 5.1.1. *Let $\mathcal{M}$ be a structure.*

1. *An amalgamation problem (for types) of length n is given by the following data:*

   (i) *A base set, A;*

   (ii) *Types $p_i(x_i)$ over A for $1 \leq i \leq n$;*

   (iii) *Types $r_{ij}(x_i, x_j)$ over A for $1 \leq i < j \leq n$;*

   subject to the conditions:

   (iv) *r_{ij} contains $p_i(x_i) \cup p_j(x_j)$;*

   (v) *$r_{ij}(x_i, x_j)$ implies the independence of x_i from x_j.*

2. *A solution to such an amalgamation problem is a type r of an independent n-tuple $x_1, \ldots, x_n$ such that the restrictions of r coincide with the given types.*

Definition 5.1.2. *A structure $\mathcal{M}$ has the type amalgamation property if whenever $(p_i; r_{ij})$ is an amalgamation problem defined over an algebraically closed base set in $\mathcal{M}^{\mathrm{eq}}$, then the amalgamation problem has a solution.*

Our goal here is to prove that Lie coordinatized structures have the type amalgamation property. By absorbing the base set A into the language we may suppose it coincides with $acl(\emptyset)$ and we will do so whenever it is notationally convenient. Our usual notation for an amalgamation problem will be either $(p_i; r_{ij})$ or just (r_{ij}), assuming the length n is known. Occasionally we will take note of generalized amalgamation problems where other restrictions are placed on the desired type r.

We build up to the general result via a series of special cases, beginning with types in a single geometry. The general result does not follow directly from the case of a single geometry, but reflects more specific properties of the geometries, as is seen in the proof of Lemma 5.1.13.

Lemma 5.1.3. *Let J be a Lie geometry, and $(p_i; r_{ij})$ an amalgamation problem of length n in which the p_i are types of sequences of elements of J over $acl(\emptyset)$. Then the amalgamation problem has a solution.*

We will leave the details to the reader, but we make a few remarks. This statement essentially comes down to the fact that inner products and quadratic forms can be prescribed arbitrarily on a basis, subject to the restrictions associated with the various types of inner product.

It may be more instructive to take note of some counterexamples to plausible strengthenings of this property. We give two examples where the solution sought is not unique, and one example of an amalgamation property incorporating a bit more data which fails to have a solution.

Example 5.1.4. *Let (V, V^*) be a polar geometry, and A an affine space over V^*. Consider independent triples (a_1, a_2, a_3) with $a_1 \in V$ and $a_2, a_3 \in A$. The relevant types r_{ij} are then determined but the type of the triple depends on the value of $(a_1, a_2 - a_3)$, which is arbitrary.*

Example 5.1.5. *In a projective space $\hat{V}$ associated with a unitary geometry V over a field K of order q^2, consider the 2-type r of a pair $\hat{x}, \hat{y}$ of independent elements of $\hat{V}$ for which $(x, y) \neq 0$ and $(x, x) = (y, y) = 0$. This defines a complete type over $acl(\emptyset)$. We consider the amalgamation problem of length 3 with all r_{ij} equal to r. For an independent triple $(\hat{x}, \hat{y}, \hat{z})$ whose restrictions realize the type r, the quantity $(x, y)(y, z)(z, x)/(y, x)(z, y)(x, z)$ is a projective invariant taking on $q + 1$ possible values α/α^σ ($\alpha \in K^*$, σ an involutory automorphism of K).*

Example 5.1.6. *We will give a generalized amalgamation problem of length 4, determined by a compatible family of 3-types r_{ijk} over $acl(\emptyset)$ of independent triples, which has no solution. Let V be a symplectic space, A affine over V, and consider the type of a quadruple x_1, x_2, x_3, x_4 with $x_1 \in V$ and the remaining x_i affine. Let the types r_{1ij} all contain the requirement: $(x_1, x_i - x_j) = 1$. These requirements are incompatible.*

Lemma 5.1.7. *Let $\mathcal{M}$ be a structure, and suppose that every amalgamation problem of length 3 in $\mathcal{M}$ over an algebraically closed subset has a solution. Then every amalgamation problem in $\mathcal{M}$ has a solution.*

Proof. This is a straightforward induction. Collapse the last two variables $x_{n-1} x_n$ to one variable y_n and define a new amalgamation problem (r'_{ij}) of length $n - 1$. The only point which requires attention is the choice of the types $r'_{i,n-1}$, which are 3-types when written in terms of the x_i. These are taken to be solutions to the type amalgamation problem $(r_{i,n-1}, r_{i,n}, r_{n-1,n})$ of length 3. ∎

In the next lemma we find it convenient to deal with a variant form of amalgamation problem incorporating some additional information.

Lemma 5.1.8. *Let M be a weakly Lie coordinatized structure, and J a geometry of M. Suppose that $(p_i; r_{1,i}, r_{2,...,n})$ is a generalized amalgamation problem over $acl(\emptyset)$ in which p_1 is the type of some element of J and $r_{2,...,n}$ is the type of an independent $(n-1)$-tuple, with the types r extending the corresponding types p appropriately. Then this generalized amalgamation problem has a solution.*

Proof. We fix a realization $(c_2, \ldots, c_n)$ of $r_{2,...,n}$, we set $C_i = acl(c_i) \cap J$, and we choose $c_1^i c_i$ satisfying r_{1i} for $2 \leq i \leq n$. We define an auxiliary generalized amalgamation problem in J by setting $r_{1i}' = tp(c_1^i C_i)$, $r_{2,...,n}' = tp(C_2, \ldots, C_n)$. By inspection of the geometries, this type of problem has a solution r'. We may choose c_1' so that $c_1' C_2 \ldots C_n$ realizes the type r'. As any c_i-definable subset of J is C_i-definable, we find that $tp(c_1' c_i) = tp(c_1 c_i)$ and the sequence $c_1, c_2, \ldots, c_n$ is independent. ∎

Roughly speaking our goal is now to treat the general amalgamation problem of length 3 by reduction to the case in which the type p_1 has rank 1. More specifically we deal with the following notion.

Definition 5.1.9. *Let M be a weakly Lie coordinatized structure and J one of its geometries.*

A semigeometric 1-type relative to J is the type over $acl(\emptyset)$ of some pair (a, b) with $a \in J$ and b algebraic over a. The multiplicity of such a type is the multiplicity of b over a.

Lemma 5.1.10. *Let M be a weakly Lie coordinatized structure and suppose that every amalgamation problem $(p_i; r_{ij})$ of length 3 with p_1 semigeometric has a solution. Then every amalgamation problem of length 3 has a solution.*

Proof. If we can solve amalgamation problems with p_1 semigeometric, then by compactness we can solve amalgamation problems in which p_1 is a type in infinitely many variables, representing the full algebraic closure in M^{eq} of an element of a geometry of M.

We now argue by induction on the rank of p_1, which we may take to be at least 1. Let c_1 realize p_1 and let $a_1 \in acl(c_1)$ belong to a coordinatizing geometry J of M. Let A be $acl(a_1)$ in M^{eq} and $p_1' = tp(A)$.

Take c_2, c_3 independent and such that $c_1 c_i$ realizes the type r_{1i} for $i = 2, 3$. Let $r_{1i}' = tp(Ac_i / acl(\emptyset))$ and $r_{23}' = r_{23}$. Then (r_{ij}') gives an amalgamation problem of length 3 of the type referred to at the outset. Let r' be a solution to this problem. We may suppose that $Ac_2 c_3$ satisfies r'.

Now we will work over A with $p_i'' = tp(c_i / A)$ for $i = 1, 2, 3$ and $r_{ij}'' = tp(c_i c_j / A)$. By the choice of r' this is an amalgamation problem, and the rank of p_1' is less than the rank of p_1, so we conclude by induction. ∎

Before treating the general amalgamation problem of length 3 with p_1 semigeometric, we will deal with the case in which $r_{12} = r_{13}$ up to a change of variable. We begin with some technical considerations.

Definition 5.1.11. *Let M be a structure, E a definable binary relation, D a definable set, and a, b elements of M.*

    1. E is a generic *equivalence relation on D if it is generically symmetric and transitive: for any independent triple a, b, c in its domain, $E(a, b)$ and $E(b, c)$ imply $E(b, a)$ and $E(a, c)$.*

    2. An indiscernible sequence I is 2-independent *if $acl(a) \cap acl(b) = acl(\emptyset)$ for $a, b \in I$ distinct.*

    3. $E_2(x, y)$ is the smallest equivalence relation containing all pairs belonging to infinite 2-independent indiscernible sequences.

Lemma 5.1.12. *Let M be $\aleph_0$-categorical of finite rank, and E a generic equivalence relation defined on the locus of a complete type p over $acl(\emptyset)$. Then*

    1. *E agrees with a definable equivalence relation E^* on independent pairs from p.*

    2. *If every pair of elements belonging to an infinite 2-independent indiscernible sequence belongs to E, then any pair of independent realizations of p belongs to E.*

Proof.

    Ad 1. Define $E^*(x, y)$ by "$p(x)$ and $p(y)$ hold and either $x = y$ or there is a z which realizes p and is independent from x, y such that $E(x, z)$ and $E(y, z)$ both hold." This is easily seen to agree with E on independent pairs, and is reflexive and symmetric. We check transitivity.

    Assume $E^*(a, b)$ and $E^*(b, c)$ hold, specifically

$$E(a, d_1), E(b, d_1), E(b, d_2), E(c, d_2)$$

with d_1 independent from a, b and d_2 independent from b, c; we may assume, in fact, that d_2 is independent from a, b, c, d_1. Then a, d_1, d_2 and b, d_1, d_2 are independent triples and thus $E(d_1, d_2)$ and $E(a, d_2)$ hold. Thus $E^*(a, c)$ holds.

    Ad 2. In view of the preceding and the hypotheses, we may assume that E is a definable equivalence relation containing E_2. It suffices now to show that any two elements of M with the same type over $acl(\emptyset)$ are E_2-equivalent. We show in fact that M/E_2 is finite, and hence is part of $acl(\emptyset)$ in M^{eq}, yielding the claim.

    Suppose toward a contradiction that M/E_2 is infinite. We will choose realizations a_i of p inductively, distinct modulo E_2, so that

$$acl(a_n) \cap \bigcup_{i < n} acl(a_i)) = acl(\emptyset).$$

Then we may suppose that the sequence $I = (a_i)$ is also indiscernible, and we have a blatant contradiction to the definition of E_2.

For the choice of a_n given a_i $(i < n)$ we first choose a new E_2-class C outside $acl(\emptyset)$ independent from $a_1, \ldots, a_{n-1}$ and then choose $a \in C$ independent from $a_1, \ldots, a_{n-1}$ over C. ∎

Lemma 5.1.13. *Let $\mathcal{M}$ be a weakly Lie coordinatized structure. Let $(p_i; r_{ij})$ be an amalgamation problem of length 3 over $acl(\emptyset)$ with p_1 semigeometric and with $r_{12} = r_{13}$ up to a change of variable; in particular $p_2 = p_3$. Then the amalgamation problem has a solution.*

Proof. As a matter of notation, take $p_1 = p_1(xy)$, $p_i = p_i(z_i)$ for $i = 2, 3$. Let J be the geometry in which the first coordinates of realizations of p_1 are found, and let C be the set defined by p_2 or p_3. We make a preliminary adjustment to ensure that for $c \in C$ we have

$$(*) \qquad\qquad r_{12}(xy, c) \text{ isolates a type over } acl(c).$$

We may replace c by some $c' \in acl(c)$ such that $c \in dcl(c')$ and $r_{12}(xy, c')$ isolates a type r'_{12} over $acl(c) = acl(c')$; the condition "$c \in dcl(c')$" means that c' can be thought of as being an extension cc'' of c. We then replace the given amalgamation problem by a problem (r'_{ij}) in which $r'_{23}(z'_1 z'_2)$ is any complete type over $acl(\emptyset)$ extending $r_{23}(z'_1 z'_2) \cup p'(z_1) \cup p'(z_2)$ where p' is the type of c' and the connection between the variables z_i and z'_i reflects the relation $c \in dcl(c')$; one may even suppose that z_i is an initial segment of z'_i. After these adjustments $(*)$ holds.

Now for $a \in J$ satisfying p_1 and $c, c' \in C$ we consider the set $B(a, c) = \{y : r_{12}(ay, c)\}$ and the sets $J(c) = \{a \in J : B(a, c) \neq \emptyset\}$ and $J(c, c') = \{a \in J : B(a, c) = B(a, c') \neq \emptyset\}$. In particular $J(c, c') \subseteq J(c) \cap J(c')$. We define a relation E on C as follows: $E(c, c')$ if and only if $J(c, c')$ is infinite. Using our understanding of J we will show that E is a generic equivalence relation extending E_2, and hence by the preceding lemma that $E(c_2, c_3)$ holds for any independent pair c_2, c_3 in C, in particular for a realization of r_{23}. This then allows us to solve the amalgamation problem directly.

We now check that E contains all pairs belonging to some infinite 2-independent indiscernible sequence I. Let μ be the multiplicity of the semigeometric type p_1 and let I' be a subset of I of cardinality 2^μ. By Lemma 5.1.8 we can find an element a independent from I' such that $B(a, c) \neq \emptyset$ for $c \in I'$. As this gives us 2^μ nonempty subsets $B(a, c)$ of $\{b : p_1(a, b)\}$, two of them must coincide, and then by indiscernibility, any two of them must coincide. As there are infinitely many such elements a, $E(c, c')$ holds for pairs in I.

It remains to be seen that E is a generic equivalence relation. We take c, c', c'' independent with $E(c, c')$ and $E(c', c'')$ holding. Thus $J(c, c')$ and $J(c', c'')$ are infinite subsets of $J(c')$, and we claim that $J(c, c'')$ is also infinite; in fact we claim that the intersection $J(c, c') \cap J(c', c'')$ is itself infinite. This involves specific features of the geometry J. We consider two representative cases: an affine space, and a linear space with a quadratic form.

Let A be an affine space corresponding to a linear model V, with V^* the definable dual. Let W_c denote the minimal $acl(c)$-definable subspace of V of finite codimension. Then $J(c)$ contains all but finitely many elements of some coset of W_c in A. Similarly, $J(c, c')$ contains all but finitely many elements of some coset of the minimal $acl(c, c')$-definable subspace $W_{c,c'}$ of finite codimension. Now $W_{c,c'} + W_{c',c''} \le W_{c'}$ is definable over both $acl(c, c')$ and $acl(c', c'')$, and as c, c', c'' are independent, this space is definable over $acl(c')$. Thus the sum equals $W_{c'}$, which means that any two cosets of $W_{c,c'}$ and $W_{c',c''}$ will intersect; the intersection is then infinite, being a coset of $W_{c,c'} \cap W_{c',c''}$. This completes the proof in the affine case.

If J is linear and carries a quadratic form then the argument is similar, but the sets involved contain almost all elements of a subset of the spaces W_c, $W_{c,c'}$ on which the quadratic form Q takes on a specific value. This set will be infinite on any subspace of J of finite codimension.   ∎

Lemma 5.1.14. *Let $\mathcal{M}$ be weakly Lie coordinatized. Let $(p_i; r_{ij})$ be an amalgamation problem of length 3 over $acl(\emptyset)$ with p_1 semigeometric. Then the problem has a solution.*

Proof. We proceed by induction on the multiplicity μ of p_1.

Take realizations $a_1 b_1 c_i$ of r_{1i} for $i = 2, 3$. If the multiplicity of b_i over $a_1 c_i$ is μ for $i = 2, 3$ then we may use Lemma 5.1.8 to choose $a_1 c_2 c_3$ appropriately, and then add b_1.

Accordingly, we may assume

The multiplicity of b_1 over $a_1 c_2$ is less than μ.

In this case the basic idea is to absorb the parameter c_2 into the base of the type and continue by induction. We first expand c_2 to an algebraically closed set C_2 and adjust the amalgamation problem accordingly. We will keep the notation as before apart from writing C_2 for c_2. The types involved now have infinitely many variables but this can be handled using the compactness theorem.

Let $C_2 c_3$ realize r_{23} and suppose $a_1 b_1 c_3$ realizes r_{13} with $a_1 b_1$ independent from from $C_2 c_3$. Take C_2' with $a_1 b_1 C_2'$ realizing r_{12} and C_2' independent from $a_1 b_1 C_2 c_3$. We will use C_2' as the basis of a new amalgamation problem.

Let $r_{13}' = tp(a_1 b_1 / C_2')$, $r_{23}' = tp(C_2 c_3 / C_2')$. To complete the specification of our auxiliary amalgamation problem, we will require a type $r_{12}'(xy, z)$ over C_2' implying the independence of xy from z and compatible with $tp(a_1 b_1 / C_2')$, $tp(C_2 / C_2')$, and $r_{12}(xy, z)$. If we construe the desired r_{12}' as a type in the variables xy, z, z', with z' replacing C_2', then this is itself an amalgamation problem involving the types $r_{12}(xy, z)$, $r_{12}(xy, z')$, and $tp(C_2, C_2')$. This case is covered by the preceding lemma. Thus we have a new amalgamation problem (r_{ij}') defined over C_2', containing the original problem. As the multiplicity of the initial 1-type $p'_1 = tp(a_1 b_1 / C'_2)$ is less than μ, we may conclude by induction. ∎

Proposition 5.1.15. *Let $\mathcal{M}$ be weakly Lie coordinatized. Then $\mathcal{M}$ has the type amalgamation property.*

The following corollary shows that the Shelah degree is bounded by the rank.

Corollary 5.1.16. *Let $\mathcal{M}$ be a weakly Lie coordinatized structure, or more generally an $\aleph_0$-categorical structure of finite rank with the type amalgamation property. Let I be an independent set, $p(x)$ a complete type over $acl(\emptyset)$, and $\varphi_a(a, x)$ ($a \in I$) a collection of formulas for which $\varphi_a \& p$ is consistent of rank $rk\, p$. Then $\bigwedge_I \varphi_a \& p$ is consistent of rank $rk\, p$.*

Proof. We may assume first that I is finite and then that $|I| = 2$, as the statement is iterable. So we consider $\varphi_1(a_1, a_3) \& \varphi_2(a_2, a_3) \& p(a_3)$, with a_1, a_2 independent. This can be converted into an amalgamation problem of the type covered by the preceding proposition. ∎

We now concern ourselves with the number of types of various sorts existing over finite sets of a given order.

Lemma 5.1.17. *Let $\mathcal{M}$ be a weakly Lie coordinatized structure, and $\varphi(x, y)$ an unstable formula. Then for each n there is a set I of size n over which there are 2^n distinct φ-types. In particular φ has the independence property.*

Proof. The instability of φ means that there is an infinite sequence I of parameters (a_i, b_i) such that $\varphi(a_i, b_j)$ will hold if and only if $i < j$. We may take I to be indiscernible. I is independent over a finite set B and we may take it to be indiscernible over B, which we absorb into the language. Let $p = tp(b_i / acl(\emptyset))$. The formulas $\varphi(a_i, x)$ and $\neg\varphi(a_i, x)$ are consistent with p and of maximal rank, so the same applies to their various conjunctions by the preceding corollary. ∎

Lemma 5.1.18. *Let $\mathcal{M}$ be Lie coordinatized with finitely many sorts, and J a 0-definable geometry of $\mathcal{M}$. Then for $X \subseteq M$ finite, and $b \in M$, we have the following estimate, uniformly:*

$$|acl(Xb) \cap J| = O(|acl(X) \cap J|).$$

Proof. Let $J(X) = acl(X) \cap J$, $J(Xb) = acl(Xb) \cap J$. It suffices to show that $\dim(J(Xb)/J(X)) = rk\, b$. As J is stably embedded with weak elimination of imaginaries, a basis B for $J(Xb)$ modulo $J(X)$ will be independent from X over $J(X)$. Thus $\dim(J(Xb)/J(X)) = rk(B/X) \leq rk(b/X) \leq rk\, b$. ∎

Lemma 5.1.19. *Let $\mathcal{M}$ be a Lie coordinatized structure with finitely many sorts, J a b-definable Lie geometry. Then for X varying over algebraically closed subsets of $\mathcal{M}$ we have*

$$|acl(Xb) \cap J| = O(|X|).$$

Proof. All cases are controlled by the projective case, so we assume that J is projective. Let J' be a canonical projective geometry nonorthogonal to J, with defining parameter $b' \in dcl(b)$.

If $b' \in acl(X)$, then $acl(Xb') \cap J' \subseteq X$ and otherwise, $acl(Xb') \cap J' = \emptyset$, so in any case $|acl(Xb') \cap J'| \leq |X|$. Thus by the previous lemma

$$|acl(Xb) \cap J| \leq |J' \cap acl(Xb)| = O(|acl(b'X) \cap J'|) = O(|X|). \quad \blacksquare$$

Proposition 5.1.20. *Let $\mathcal{M}$ be Lie coordinatizable, $D \subseteq \mathcal{M}$ 0-definable of rank k. Then the number of types of elements of D over an algebraically closed set of order n in $\mathcal{M}$ is $O(n^k)$.*

Proof. Suppose first that $D = J$ is a coordinatizing geometry of $\mathcal{M}$. For algebraically closed X the types under consideration are determined by their restrictions to $X \cap J$. Thus we may assume $\mathcal{M} = J$ in this case. The statement is then clear by inspection. For example, in the presence of a quadratic form, the behavior of the the form on an extension of a subspace by a single point is determined by its value on the additional point and an induced linear function defined on the subspace. If the geometry is affine the situation remains much the same.

We turn to the general case. We may assume that D is the locus of a single type. Take $c \in D$ of rank k and $b \in acl(c)$ of rank $k-1$ supporting a coordinate geometry J_b, with $a \in J_b$ such that $c \in acl(ba)$. Let D', D'', and D''' be the loci of the types of b, ba, and bac respectively. Inductively, the number of types of elements of D' over an algebraically closed subset X of order n is $O(n^{k-1})$. By Lemma 5.1.19 for $b \in D'$ we have $|acl(Xb) \cap J| = O(|X|)$ and thus the number of types in J over $acl(Xb)$ is also $O(|X|)$. Thus the number of types in D'' over X is $O(n^k)$. As D''' is a finite cover of D'' the number of types of elements in D''' is also $O(n^k)$ and as the types of elements of D lift to types of elements of D''' this bound applies to D'''. $\blacksquare$

Definition 5.1.21. *For D a definable set let $s(D, n)$ denote the minimum number of types of elements of D existing over a subset of D of order n.*

Observe, for example, that in one of the standard geometries this will be $O(n)$, with the optimal subset being as close to a subspace as possible.

The following corollary depends on estimates for the sizes of envelopes to be given shortly.

Corollary 5.1.22. *Let M be Lie coordinatized with finitely many sorts, D a 0-definable subset of M. Then $s(D, n)$ is polynomially bounded.*

Proof. We show in Proposition 5.2.2 below that the size of D in an envelope E is given by a polynomial function of certain quantities q^d, q being approximately the size of the base field and d varying over the dimensions of E. Varying just one of these dimensions, we can find envelopes in which the size of D is asymptotically a constant times q^d for some d. Thus for m large we can find envelopes E in which the size of D is comparable to m; that is, $m \leq |D| \leq (q + \epsilon)m$. Thus taking X to be a subset of $D \cap E$ of order m and applying the previous result, we get the desired bound.   ∎

We mention two problems. The first relates to the amalgamation of types.

Problem 1. *Find independent elements a_1, a_2, a_3 such that there is no B independent from $a_1a_2a_3$ for which:*

$$tp(a_1a_2/B) \cup tp(a_1a_3/B) \cup tp(a_2a_3/B) \text{ determines } tp(a_1a_2a_3/B).$$

Problem 2. *Are types over envelopes uniformly definable?*

5.2 THE SIZES OF ENVELOPES

We deal here with the computation of the size of an envelope as a function of its dimensions, and also with the sizes of the automorphism groups. We wish to express the sizes of envelopes as polynomial functions of the relevant data, and to do so it will be convenient to work with square roots of the sizes of the associated fields.

Notation 5.2.1. *Let M be Lie coordinatized and p a canonical projective geometry. For an envelope E we let $d_E(p)$ be the corresponding dimension (or cardinality in the degenerate case) and we let $d_E^*(p) = (-\sqrt{q})^{d_E(p)}$, where q is the size of the base field; in the degenerate case we set $d^*(p) = \sqrt{d(p)}$. When E is understood we write $d(p)$ and $d^*(p)$.*

Proposition 5.2.2. *Let $\mathcal{E}$ be a family of envelopes for the Lie coordinatized structure M such that for each dimension p corresponding to an orthogonal space, the signature and the parity of the dimension is constant on the family. Then there is a polynomial ρ in several variables such that for every E in $\mathcal{E}$, $|E| = \rho(d^*(E))$, where $d^*(E)$ is the vector $(d_E^*(p))$. The total degree of ρ is $2\,rk(M)$ and all leading coefficients are positive. If M is the locus of a single type (with the coordinatization in M^{eq}), then ρ is a product of polynomials in one variable.*

Proof. We show that for any definable set D_a of M, there is a polynomial of the type described giving the cardinality of D_a in any $E \in \mathcal{E}$ which contains the parameter a. We may suppose that D_a is the locus of a single type over a. We will proceed by induction on $rk(D_a)$.

Take $d \in D_a$ and $c \in acl(ad)$ lying in an a-definable geometry J, which we may take to be degenerate, linear, or affine, with associated canonical projective p. Let D'_{ac} be the set of realizations of $tp(d/ac)$. Then we may take $\rho_{D_a} = \rho_J \rho'_{D_{ac}} / Mult(c/ad)$. This reduces to the case $D = J$.

If J is affine or quadratic, add a parameter to reduce to a basic linear geometry J. Then the dimension of J in E is $d_E(p)$ minus a constant depending on the type of a. Thus it suffices to find a polynomial giving the number of realizations of a type in J in terms of $d_E^*(p)$ or equivalently in terms of the corresponding expression $(\pm\sqrt{q})^{\dim J}$. The essential point is to compute the sizes of sets defined by equations $Q(x) = \alpha$ with Q a quadratic or unitary form. Let $n(d, \alpha)$ be this cardinality as a function of the dimension and α, depending also the type of the geometry. These are straightforward computations. We give details.

In the orthogonal case we can break up the space as the orthogonal sum of a $2i$-dimensional space H with a standard form $Q(\bar{\alpha}, \bar{\beta}) = \sum \alpha_i \beta_i$ and a complement of dimension $j \leq 2$. So on H we have $n(2i, 0) = (q^i - 1)q^{i-1} + q^i$ and $n(2i, \alpha) = (q^{2i} - n(2i, 0))/(q - 1)$ for $\alpha \neq 0$. Thus on the whole space

$$n(2i + j, \alpha) = n(2i, 0)n(j, \alpha) + [(q^{2i} - n(2i, 0)/(q - 1)](q^j - n(j, \alpha))$$

where the parameter n is computed with respect to the corresponding induced form. This simplifies to

$$n(2i + j, \alpha) = q^i n(i, \alpha) + q^{j-1}(q^{2i} - q^i)$$

and for small i $n(i, \alpha)$ is treated as a constant, corresponding to the particular form used.

In the unitary case $n(d, \alpha)$ is independent of α for α nonzero and thus it suffices to compute $n(d, 0)$. Using an orthonormal basis and proceeding inductively one gets $n(d, 0) = q^{d-1}(\sqrt{q} + 1) - n(d - 1, 0)\sqrt{q}$ and then $n(d, 0) = q^d/\sqrt{q} + (-\sqrt{q})^{d-1}(1 - \sqrt{q})$. ∎

Remark 5.2.3. If we are working with graphs, for example, the number of edges is given by a polynomial. The polynomials ρ can be determined given a sufficiently large envelope in which the subenvelopes are known.

We now discuss the chief factors of automorphism group of an envelope, which are the successive quotients in a maximal chain of normal subgroups of this group.

Lemma 5.2.4. *Let G be the automorphism group of the envelope $E(d)$ in a Lie coordinatized structure $\mathcal{M}$. Then the number of chief factors of G is bounded, independently of d, and each chief factor is of one of the following kinds:*

1. *abelian;*
2. $H^{\rho(d)}$, *where H is a fixed finite group and ρ is one of the functions described in the preceding proposition;*
3. $K^{\rho(d)}$, *with $\rho(d)$ as in the preceding proposition and K a classical group $PSL(d_i, q_i)$, $PSp(d_i, q_i)$, $P\Omega^{\pm}(d_i, q_i)$, $PSU(d_i, q_i)$, or $Alt(d_i)$, as appropriate to the ith dimension.*

Proof. Once the dimensions are sufficiently large, the socle of the automorphism group of one layer of the coordinate tree over the previous layer is of the form (3) or abelian, unless the geometry is finite (in $\mathcal{M}$), with the number of factors corresponding to the size of a definable set modulo an equivalence relation. The remainder of the automorphism group at that layer is solvable. If the layer consists of copies of a finite geometry, consider a chief factor H/K with H, K $Aut(E)$-invariant subgroups acting trivially on the previous layer. Let A be the automorphism group of the finite geometry involved, and let L be the part of E lying in the previous level of the coordinate tree, so that H, K lie in A^L. If H/K is nonabelian then it is a product of a certain number of copies of a single isomorphism type of finite simple group S. The number of factors is the order of L modulo the following equivalence relation: $a \sim b$ if the projection of H/K onto $A_a \times A_b$ is a diagonal subgroup isomorphic to S. This relation is $Aut(E)$-invariant and hence definable. Thus the number ρ of factors involved is equal to the size of a definable set in an envelope (a definable quotient of L). ∎

Corollary 5.2.5. *Let $\mathcal{M}$ be a Lie coordinatized structure. Then for the dimension function d large enough, $Aut(E(d))$ determines d up to a permutation of the coordinates and up to orientation in the odd-dimensional orthogonal case.*

Proof. Let f be a bound on the size of the chief factors of the second type above. Let d be large enough that the chief factors of the third type are all of order greater than f. Then these chief factors can be recovered from the automorphism group unambiguously and the data d can be read off. ∎

Lemma 5.2.6. *Let $\mathcal{M}$ be a Lie coordinatized structure and D a definable subset. Then the following are equivalent:*

1. $rk(D) < rk(\mathcal{M})$.
2. $\lim_{E \to \mathcal{M}} |D[E]|/|E|) = 0$.

Here the limit is taken over envelopes whose dimensions all go to infinity, and $D[E]$ means D taken in E, which for large enough E is $D \cap E$. The convergence is exponentially rapid if all geometries are nondegenerate.

Proof. We compare the polynomials ρ_D, ρ_E giving the sizes of D and E.

If the ranks are equal, then both polynomials have positive leading coefficients and total degree $2\,rk(\mathcal{M})$. For each dimension d_i, ρ_D, ρ_E involve the parameter $d_i^* = \alpha_i^{d_i}$ for an appropriate α_i (read this expression as d_i in the degenerate case). Let the dimensions d_i be taken momentarily as arbitrary real numbers going jointly to infinity along the curve $d_1^* = d_2^* = \ldots$, so that the polynomials ρ_D, ρ_E reduce to one variable polynomials converging to a positive γ. After a slight perturbation we may suppose that $d_1, d_2, \ldots$ are rational, that ρ_D/ρ_E approaches γ, and that the terms of total degree less than $2\,rk(\mathcal{M})$ make a negligible contribution. After rescaling by a common denominator, the "dimensions" are integers, the ratio of the highest order parts of ρ_D and ρ_E goes to γ, and the lower-order terms are even more negligible. Thus we have a sequence of dimension assignments tending jointly to infinity on which the quotient ρ_D/ρ_E will not go to zero.

Now assume that $rk(D) < rk(\mathcal{M})$. We may take D, E to be realizations of single types, so that ρ_D and ρ_E factor as products of polynomials in one variable $\rho_{D,i}$, $\rho_{E,i}$. The ratios $\rho_{D,i}/\rho_{E,i}$ are bounded, as otherwise varying only the one relevant dimension we would get a proper subset with more elements than the whole set E. On the other hand at least one of the $\rho_{D,i}$ has degree less than the degree of $\rho_{E,i}$ so the limit goes to 0 (rapidly, if the geometry is nondegenerate). ∎

We now prove a finitary Löwenheim–Skolem principle.

Lemma 5.2.7. *Let $\mathcal{M}$ be Lie coordinatized. For any subset X of $\mathcal{M}$ there is an envelope E of $\mathcal{M}$ containing X, in which each dimension is at most $2\,rk(X) \le 2\,rk(\mathcal{M}) \cdot |X|$.*

Proof. Let $J_1, \ldots, J_n$ be the $acl(\emptyset)$-definable dimensions, and $E_i = acl(X) \cap J_i$. The dimension of E_i is at most $rk(X)$. If the geometry J_i carries a form then increase E_i to a nondegenerate subspace, of dimension at most $2\,rk(X)$. Let $\mathcal{M}'$ be a maximal algebraically closed subset of $\mathcal{M}$ containing X, and such that $\mathcal{M}' \cap J_i = E_i$. Then $\mathcal{M}'$ is Lie coordinatized and has smaller rank, unless these geometries are finite, in which case iteration of the process will eventually lower the rank or the height of the coordinatizing tree. By induction on rank we may suppose that in $\mathcal{M}'$ there is an envelope E with the desired properties. This will then be an envelope in $\mathcal{M}$, with the desired properties. ∎

Remark 5.2.8. The existence of indiscernible sets of order n in all large finite structures with a fixed number of 5-types is proved in [CL]. In particular, an infinite quasifinite structure contains an infinite set of indiscernibles. Conversely, from the latter result it follows that there is a constant c such that for large n, a pseudofinite structure with at least c^n elements contains a sequence of indiscernibles of length n. This follows from the last lemma using the bounds on the sizes of envelopes, since the ranks involved can be bounded in terms of the number of 4-types. It is possible that an explicit bound of this kind can also be extracted by tracing through the arguments in [CL].

Problem 3. *Do the abelian chief factors of automorphism groups of envelopes have orders $p^{\sigma(d,d^*)}$ with σ a polynomial similar to ρ—in particular, a product of polynomials in one variable (i.e., depending on one dimension)?*

One can treat the case of affine covers by dualization, reducing to finite covers. Then by results in [EH] the problem reduces to the following: if J is a definable combinatorial geometry on a definable set D of a Lie structure $\mathcal{M}$, which is subordinate to algebraic closure, show that the dimension of J in an envelope of $\mathcal{M}$ is given by a polynomial in d, d^*.

5.3 NONMULTIDIMENSIONAL EXPANSIONS

We show here that Lie coordinatizable structures have "nonmultidimensional" expansions, lifting [HrTC, §3] to the present context. As in that earlier case, the difficulty lies in the interaction of orthogonal geometries, which means that the outer automorphism groups may be related even if the simple parts of the groups are not.

Definition 5.3.1. *A Lie coordinatized structure is said to be* nonmultidimensional *if it has only finitely many dimensions, or equivalently (and more explicitly) if all canonical projectives are definable over $acl(\emptyset)$.*

Proposition 5.3.2. *Every Lie coordinatized structure can be expanded to a nonmultidimensional Lie coordinatized structure.*

Proof. We use a locally transitive coordinatizing tree, meaning that the type of a point at a given level depends only on the level. We also allow the introduction of a finite number of additional sorts, each carrying a single basic geometry.

Let M_i be the coordinatizing tree up to level i together with the elements of the special sorts, and let Δ be the set of indices i for which the geometries J_a associated to points at level i are orthogonal to M_i. We proceed by induction on M_i, the case $\Delta = \emptyset$ being the nonmultidimensional case. So we take Δ nonempty.

Now let $n \in \Delta$ be maximal. Let T_n be the set of elements lying at level n in the coordinatizing tree. For $a \in T_n$ let $P_{a'}$ be the canonical projective geometry associated with P_a and let q be the type of a'. Let $V_{a'}$ be the corresponding linear geometry. If these linear geometries are not actually present in the structure, we may attach them freely to the canonical projectives. (In the degenerate case, the geometry is considered to be both linear and projective.)

The isomorphism type of $V_{a'}$ is independent of a', but there will not be any system of identifications present between the various $V_{a'}$.

Suppose for definiteness that $V_{a'}$ is of orthogonal type in odd characteristic, with base field $K_{a'}$, and bilinear form $B_{a'} : V_{a'} \times V_{a'} \to L_{a'}$, a 1-dimensional $K_{a'}$-space. Fix a copy K of the base field, and a 1-dimensional space L over K. Fix a 2-dimensional space $U_\circ$ over K and a nondegenerate bilinear form $(\ )_\circ : U_\circ \times U_\circ \to L$ which takes the value 0 at some nonzero point. The pair $(U_\circ, (\ )_\circ)$ is unique up to an isomorphism fixing K and L.

Now let U_1, Q_1 be an infinite dimensional nondegenerate orthogonal space over the prime field $F \leq K$ and set $U = U_1 \otimes U_\circ$ as a K-space. The forms $(,)_\circ$ and $(,)_1$ induce a bilinear form $(,)$ on U satisfying $(a_1 \otimes a_\circ, b_1 \otimes b_\circ) = (a_1, b_1)_1 \cdot (a_\circ, b_\circ)_\circ$. This makes sense by the universal property of tensor products. Let Γ be the family $\{a \otimes U_\circ : a \in U_1\}$. Then

(1) Any automorphism h of (K, L) extends to an automorphism of U fixing Γ pointwise.

The uniqueness of $U_\circ$ signifies that h extends to $U_\circ$. To extend to U fix U_1 pointwise. Then Γ is fixed pointwise.

Add U as a new sort. For b satisfying q pick isomorphisms $h_b : U \to V_b$, and let $\Gamma_b = h_b[\Gamma]$. Let $\mathcal{M}'$ be $\mathcal{M}$ expanded by the sort U and a family of maps $f_b : \Gamma \to \Gamma_b$ for b satisfying q. f_b is to be coded by a ternary relation on $q \times U \times \bigcup_b V_b$. h_b is not part of the structure but the sets Γ and Γ_b can be recovered from f_b in $(\mathcal{M}')^{\text{eq}}$. We claim that $\mathcal{M}'$ remains 4-quasifinite and that Δ is reduced by 1.

By a *normal* subset of $\mathcal{M}^{\text{eq}}$ we mean a union of 0-definable sets. The restriction of a normal subset to a finite number of sorts is then 0-definable. We consider normal subsets S satisfying the additional condition:

For b satisfying q, V_b is orthogonal to S.

This means that any basic geometry corresponding to V_b (with $acl(b)$ fixed) is orthogonal to S. Let Q be a maximal normal subset of this type containing T_n. Then Q contains the locus of q and is algebraically closed. We claim that Q is also stably embedded in $\mathcal{M}$, since for any projective or affine geometry in Q, if the dual exists in $\mathcal{M}$, then it is contained in Q.

We claim now:

(2) For any automorphisms α of Q and β of U, the map $\alpha \cup \beta$ is induced by an automorphism of $\mathcal{M}'$.

Let $Q_1 = Q \cup \bigcup_b V_b$. Then Q_1, like Q, is stably embedded in $\mathcal{M}$. We first extend $\alpha \cup \beta$ to Q_1. For b satisfying q, α induces maps $K_b \to K_{\sigma b}$ and L_b to $L_{\sigma b}$. By (1) these maps are induced by a linear isomorphism $\theta_b : V_b \to V_{\sigma b}$ compatible with $f_{\sigma b} \beta f_b^{-1}$. Using the orthogonality condition, $\alpha \cup \beta \cup \bigcup_b \theta_b$ is elementary and extends to an automorphism of $\mathcal{M}'$.

It remains to be seen that apart from the introduction of U, the rest of the coordinatization of $\mathcal{M}$ is unaffected; specifically, if J_c is a canonical projective geometry of $\mathcal{M}$ orthogonal to the geometries V_b, then

> J_c has no extra structure as a subset of $\mathcal{M}'$;
> If J_c is stably embedded in $\mathcal{M}$, then it remains stably embedded in $\mathcal{M}'$.

We may assume that J_c is stably embedded in $\mathcal{M}$. If J_c is contained in Q this follows from (2), and otherwise any automorphism of J_c fixing $acl(c)$ extends to an automorphism of $\mathcal{M}$ fixing Q_1 pointwise. This is then elementary in $\mathcal{N}'$.

This completes the orthogonal case in odd characteristic. The linear, symplectic, and unitary cases are similar, with the auxiliary space $U_\circ$ 1-dimensional in the unitary case. In the orthogonal case in characteristic 2, the orthogonal geometry is an enrichment of a symplectic geometry and we may suppose that the pure symplectic space occurs as well, and that the quadratic form used occurs also as a point in an associated quadratic geometry. Then we can switch to the symplectic case. Similarly, in the case of a polar geometry (V, V^*) reduce the scalars to the prime field and introduce linear isomorphisms $\iota_V : V \to V^*$. This can be done without destroying outer automorphisms and brings us back to the symplectic case.   ∎

Proposition 5.3.3. *For $\mathcal{M}$ quasifinite the following are equivalent:*

1. *$\mathcal{M}$ is stable.*
2. *$\mathcal{M}$ is $\aleph_0$-stable.*
3. *$\mathcal{M}$ does not interpret a polar space.*

Proof. We must show that (3) implies (2). So assume (3). In particular none of the canonical geometries for $\mathcal{M}$ involve bilinear forms. The geometries occurring are therefore all strongly minimal and stably embedded. Morley rank is subadditive in the $\aleph_0$-categorical setting, for stably embedded definable subsets (cf. [HrTC]); hence, using the coordinatization, $\mathcal{M}$ has finite Morley rank.   ∎

Remarks 5.3.4

As the class of *stable* polar spaces is the class of *finite* polar spaces, which is not an elementary class, the notion of a stable quasifinite structure in a given language is not an elementary notion. On the other hand, for a fixed finite language L, the class of stable homogeneous L-structures is elementary [CL]. This can be seen fairly directly as follows. By a result of Macpherson [Mp1] in a finitely homogeneous structure, no infinite group is interpretable. In particular for finitely homogeneous structures, quasifiniteness and stability are equivalent. But for finitely homogeneous structures quasifiniteness is elementary.

Although we work outside the stable context, we still require the analysis of [CL] for primitive groups with nonabelian socle, which enters via [KLM].

5.4 CANONICAL BASES

We do not have a theory of canonical bases as such, but the following result serves as a partial substitute.

Proposition 5.4.1. *Let $\mathcal{M}$ be $\aleph_0$-categorical of finite rank, and suppose that a_1, a_2, a_3 is a triple of elements which are independent over a_1, over a_2, and over a_3. Then a_1, a_2, a_3 are independent over the intersection of $acl(a_i)$, $i = 1, 2, 3$, in $\mathcal{M}^{\mathrm{eq}}$.*

We begin with a few lemmas.

Lemma 5.4.2. *Let $\mathcal{M}$ be $\aleph_0$-categorical of finite rank and let R be a 0-definable symmetric binary relation satisfying*

> *Whenever $R(a, b)$, $R(b, c)$ hold with a, c independent over b, then $R(a, c)$ holds and b, c are independent over a.*

Then there is a 0-definable equivalence relation E such that

> *$R(a, b)$ implies the following: $E(a, b)$ holds and a, b are independent over $a/E = b/E$.*

Proof. We define $E(a, b)$ as follows: For some c independent from a over b and from b over a, $R(a, c)$ and $R(b, c)$ holds.

We check first that R implies E. If $R(a, b)$ holds, choose c independent from a over b such that $R(c, b)$ holds. Then by $(*)$ $R(a, c)$ holds and c is independent from b over a. Thus E holds. The domain of E is the same as the domain of R and E is clearly reflexive and symmetric on this domain. We now check transitivity.

Suppose $E(a_1, a_2)$ and $E(a_2, a_3)$ hold and let a_{12}, a_{23} be witnesses. Thus we have $R(a_i, a_{ij})$; $R(a_j, a_{ij})$; and a_{ij} is independent from a_i over a_j and from a_j over a_i. As a_{12} is independent from a_1 over a_2, we may take it independent from $a_1 a_2 a_3$ over a_2; and similarly for a_{23}. Furthermore, we may take a_{12}, a_{23} independent over a_1, a_2, a_3 and hence over a_2. From $R(a_2, a_{12})$ and $R(a_2, a_{23})$ we then deduce $R(a_{12}, a_{23})$.

Pick c independent from $a_1 a_2 a_3 a_{23}$ over a_{12} such that $R(a_{12}, c)$ holds. We claim then:

(1) $R(a_i, c)$ holds for all i, and
             c is independent from a_{ij} over a_i and over a_j.

First, since c is independent from a_{23} over a_{12} we get $R(a_{23}, c)$ and c is independent from a_{12} over a_{23}; the latter implies that c is independent from

$a_1a_2a_3a_{12}$ over a_{23}. So c is independent from a_1 or a_2 over c_1, and from a_2 or a_3 over c_2. By another application of $(*)$ the relation (1) follows.

Now using (1) we get c independent from $a_1a_2a_3a_{12}a_{23}$ over each a_i and, in particular, c is independent from a_3 over a_1 and from a_1 over a_3; so $E(a_1, a_3)$ is witnessed by c. Thus E is transitive.

Finally, we must show that if $R(a, b)$ holds and $c = a/E = b/E$, then a, b are independent over c. Let a' realize the type of a over c with a' independent from a over c. We will show then that a and b are independent over a' and thus a and b are independent over c.

As $E(a, a')$ holds, there is d satisfying

$R(a, d)$, $R(a', d)$, and d is independent from a over a' and from a' over a.

We will take a', d independent from b over a. In particular we have a' independent from b over ad, and b independent from d over a; the latter, with $(*)$, gives b independent from a over d and then combined with the former, we get aa' independent from b over d, hence a independent from b over $a'd$. As a is independent from d over c we get finally a independent from b over a'. ∎

Definition 5.4.3. *Let $a_1, \ldots, a_n$ be a sequence of elements in a structure of finite rank.*

1. *The sequence is said to be 1-locally independent if it is independent over any of its elements.*
2. *We set $\delta(a_1, \ldots, a_n) = \sum_i rk\, a_i - rk(a_1 \ldots a_n)$.*

Lemma 5.4.4. *Let $\mathcal{M}$ be a structure of finite rank, $\mathbf{a} = a_1, \ldots, a_n$ a sequence of elements. Then the sequence $\mathbf{a}$ is 1-locally independent if and only if:*

The quantity $\delta = \delta(a_i a_j)$ is independent of i, j (distinct);
and $\delta(\mathbf{a}) = (n - 1)\delta$.

Proof. We have in general for any fixed index k, writing $\sum'$ for a sum excluding the index k:

$$
\begin{aligned}
\delta(\mathbf{a}) &= \sum_i rk(a_i) - (rk(\mathbf{a}/a_k) + rk(a_k)) \\
&= \sum_i{}' rk(a_i) - rk(\mathbf{a}/a_k) \\
&\geq \sum{}' rk(a_i) - \sum{}' rk(a_i/a_k) = \sum{}' \delta(a_i, a_k)
\end{aligned}
$$

with equality if and only if $\mathbf{a}$ is independent over a_k. Thus if $\delta = \delta(a_i, a_j)$ is constant and $\delta(\mathbf{a}) = (n-1)\delta$, then we have equality regardless of the choice of

k and the sequence is 1-locally independent, while, conversely, if the sequence is 1-locally independent, then $\delta(\mathbf{a}) = \sum' \delta(a_i a_k)$ for any k and it suffices to check that the $\delta(a_i a_j)$ are independent of i, j. But the restriction of a to any three terms $a_i, a_{i'}, a_{i''}$ remains 1-locally independent, and applying our equation to a sequence of length 3 with $k = i'$ or $k = i''$ yields $\delta(i, i') = \delta(i, i'')$, from which it follows that δ is constant. ■

Lemma 5.4.5. *Let $\mathcal{M}$ be a structure of finite rank.*

1. *Suppose that* $\mathbf{a} = a_1, a_2, a_3, a_4$ *is a sequence such that both of the triples* a_1, a_2, a_3 *and* a_2, a_3, a_4 *are 1-locally independent. If* a_1 *and* a_4 *are independent over* $a_2 a_3$, *then* $\mathbf{a}$ *is 1-locally independent.*
2. *If* $\mathbf{a} = a_1 a_2 b_1 b_2 c_1 c_2$ *is a sequence whose first four and last four terms are 1-locally independent, and* $a_1 a_2$ *is independent from* $c_1 c_2$ *over* $b_1 b_2$, *then* $\mathbf{a}$ *is 1-locally independent.*

Proof.

Ad 1. We have $\delta(a_i a_j) = \delta$ constant, with the possible exception of the pair a_1, a_4. Repeating the calculation of the previous lemma over $a_2 a_3$ rather than a_k, using $rk(a_1 a_2 a_3 a_4 / a_2 a_3) = rk(a_1 / a_2 a_3) + rk(a_4 / a_2 a_3)$, we get $\delta(\mathbf{a}) = 3\delta$. Thus it remains only to be checked that $\delta(a_1 a_4) = \delta$. We may show easily that $\mathbf{a}$ is independent over a_2 or over a_3, starting from the independence of $a_1 a_2 a_3$ from a_4 over $a_2 a_3$. Thus

$$
\begin{aligned}
rk\, a_2 - \delta &= rk(a_2/a_1) \geq rk(a_2/a_1 a_4) \geq rk(a_2/a_1 a_3 a_4) \\
&= rk(a_2/a_3) = rk(a_2) - \delta
\end{aligned}
$$

and, in particular, we have the equation $rk(a_2/a_1 a_4) = rk(a_2) - \delta$. Now

$$
\begin{aligned}
rk(\mathbf{a}) &= rk(a_1 a_4) + rk(a_2/a_1 a_4) + rk(a_3/a_1 a_2 a_4) \\
&= rk(a_1 a_4) + (rk(a_2) - \delta) + rk(a_3) - \delta
\end{aligned}
$$

and thus

$$
3\delta = \sum rk(a_i) - rk(\mathbf{a}) = \delta(a_1 a_4) + 2\delta
$$

and $\delta(a_1 a_4) = \delta$.

Ad 2. It is straightforward that $\mathbf{a}$ is independent over b_1 or over b_2 and by symmetry it will be sufficient to prove that $\mathbf{a}$ is independent over a_1.

We have by assumption $c_1 c_2$ independent from $a_1 a_2 b_1 b_2$ over $b_1 b_2$ and thus c_1 is independent from $a_1 a_2 b_1 b_2$ over $b_1 b_2 c_2$, but also c_1 is assumed independent from $b_1 b_2 c_2$ over c_2, and thus

$$
c_1 \text{ is independent from } a_1 a_2 b_1 b_2 c_2 \text{ over } c_2.
$$

In particular, $c_1 c_2$ is independent from $a_1 b_1 b_2$ over $a_1 c_2$. By Case 1 $a_1 b_1 b_2 c_2$ is 1-locally independent and is, in particular, independent over a_1, so from the previous relation we derive the independence of $c_1 c_2$ from $b_1 b_2$ over a_1. Combining this with the independence of $c_1 c_2$ from $a_1 a_2 b_1 b_2$ over $b_1 b_2$, we find that $c_1 c_2$ is independent from $a_1 a_2 b_1 b_2$ over a_1. Now c_1 is independent from c_2 over $b_1 b_2$ and $c_1 c_2$ is independent from a_1 over $b_1 b_2$ so c_1 is independent from c_2 over $a_1 b_1 b_2$, and hence, by transitivity, over a_1. Thus $a_1 a_2 b_1 b_2$ is independent over a_1, $c_1 c_2$ is independent from $a_1 a_2 b_1 b_2$ over a_1, and c_1 is independent from c_2 over a_1. Thus **a** is independent over a_1. ∎

Proof of Proposition 5.4.1. We have a_1, a_2, a_3 1-locally independent. Let X be the set of pairs $x = (x_1, x_2)$ such that each coordinate x_1 or x_2 realizes the type of one of the three elements a_i, and define a relation R on X by: $R(x, y)$ if and only if x_1, x_2, y_1, y_2 is a 1-locally independent quadruple. We will apply Lemma 5.4.2 to R. Note first that if $R(x, y)$ and $R(y, z)$ hold with x and z independent over y then the 6-tuple (x, y, z) satisfies the conditions of case 2 of the previous lemma, and thus the six coordinates form a 1-locally independent sequence. Thus Lemma 1 applies and there is a 0-definable equivalence relation E such that

$$R(x, y) \text{ implies: } E(x, y), \text{ and } x, y \text{ are independent over } x/E.$$

Now consider the 1-locally independent triple (a_1, a_2, a_3). We extend this triple by two further elements a_4, a_5 satisfying the following three conditions: $tp(a_i/a_2 a_3) = tp(a_1/a_2 a_3)$, for $i = 4, 5$; a_4 independent from a_1 over $a_2 a_3$; and a_5 is independent from a_1, a_4 over a_2, a_3. We claim that any 4-tuple from a_1, a_2, a_3, a_4, a_5 is 1-locally independent. This follows from Lemma 5.4.5, part (1), for $a_1 a_2 a_3 a_4$, $a_1 a_2 a_3 a_5$, or $a_2 a_3 a_4 a_5$. In the remaining two cases, $a_1 a_2 a_4 a_5$ and $a_1 a_3 a_4 a_5$, we need to check that a_5 is independent from a_4 over $a_1 a_2$ or $a_1 a_3$. But a_5 is independent from a_4 over $a_1 a_2 a_3$ and from $a_1 a_2 a_3$ over a_2 or a_3. Thus all of these 4-tuples are 1-locally independent, and hence any two disjoint pairs are E-equivalent; and by transitivity any two pairs are E-equivalent. Let e be the common E-class of these pairs. Then $a_1 a_2$ is independent from $a_3 a_4$ over e and $a_1 a_3$ is independent from $a_2 a_4$ over e. In particular, working over e we have a_3 independent from $a_1 a_2$, and a_1 independent from a_2, and thus $a_1 a_2 a_3$ is an independent set over e. It remains only to be checked that e is algebraic over each a_i. Certainly $e \in acl(a_1 a_2)$ and $acl(a_3 a_4)$, and as these pairs are independent over any a_i, we have $e \in acl(a_i)$ for all i. ∎

5.5 MODULARITY

Definition 5.5.1. *Let M be $\aleph_0$-categorical of finite rank. M is modular if whenever A_1, A_2 are algebraically closed sets in M^{eq}, they are independent over their intersection.*

By convention *acl* will always be taken to operate in M^{eq}. This point may be reemphasized occasionally.

Modularity, as defined here, is called "local modularity" in the literature dealing with the case of finite Morley rank, where the term "modular" is applied only to strongly minimal sets D which in addition to the stated property have "geometric elimination of imaginaries": for $a \in D^{eq}$, there is $A \subseteq D$ with $acl(e) = acl(A)$.

As a matter of notation we will use the symbol $\perp$ for *independence*, a symbol which is more often used for model theoretic *orthogonality*; but the latter concept does not really call for any special notation in our present development.

Lemma 5.5.2. *Let M be $\aleph_0$-categorical of finite rank. Then M is modular if and only if the lattice of algebraically closed subsets of M^{eq} satisfies the modular law:*

$$a \wedge (b \vee c) = b \vee (a \wedge c) \qquad \text{for } b \leq a.$$

Proof. Suppose M is modular, and A, B, C are algebraically closed subsets of M^{eq} with $B \subseteq A$. Our claim is

$$A \cap (acl(BC)) = acl(B \cup (A \cap C))$$

the modular law. From modularity applied to A, C, as $B \subseteq A$ we deduce easily that $A \perp BC$ over $B \cup (A \cap C)$. Thus $A \cap acl(BC) = acl(B \cup (A \cap C))$.

In the converse direction, assume the modular law in M^{eq}, but A, B are algebraically closed and dependent over their intersection. Minimize $rk(A/B)$ and, subject to this constraint, $rk(A)$. We may suppose $A \cap B = acl(\emptyset)$, as the modular law holds in the corresponding sublattice (i.e., above $A \cap B$). We adopt the notation $0 = acl(\emptyset)$ for the present. After these reductions, we claim that A is a lattice atom: a minimal nontrivial algebraically closed set.

Suppose $0 < A' \leq A$ with A' algebraically closed. As $A' > A \cap B$, $rk(A'/B)$ is positive and $rk(A/A'B) < rk(A/B)$, so by minimality

$$A \perp A'B \text{ over } A \cap acl(A'B).$$

If $A \cap acl(AB')$ is independent from B over $A \cap acl(AB') \cap B = 0$, then $A \perp B$ over 0, a contradiction. Thus A may be replaced by $A \cap acl(A'B)$, and by the minimality of $rk\, A$ we find $A \subseteq acl(A'B)$. By the modular law

$$A = A \cap acl(A'B) = acl(A' \cup (A \cap B)) = A'$$

as claimed.

Now consider a conjugate B' of B over A independent from B over A. Note that

$$acl(AB) \cap B' = 0$$

since $acl(AB) \cap B' \subseteq A \cap B' = 0$. If the triple A, B, B' is 1-locally independent, then it is independent over the intersection 0 by Proposition 5.4.1, a contradiction. If it is not 1-locally independent, then either A, B are dependent over B', or A, B' are dependent over B, and in any case $rk(A/BB') < rk(A/B)$. Thus by the minimality of $rk(A/B)$, we have independence of A from BB' over $A_o = A \cap acl(BB')$. As A is an atom, we have either $A_o = 0$, contradicting the choice of A, or $A \subseteq acl(BB')$. In the latter case, applying the modular law to $acl(A, B)$, B, and B' we get $A \subseteq acl(AB) \cap acl(BB') = acl(B, acl(AB) \cap B') = B$, which is absurd. ∎

Proposition 5.5.3. *Let $\mathcal{M}$ be $\aleph_0$-categorical of finite rank. Then the following are equivalent.*

1. *$\mathcal{M}$ is modular.*
2. *For all finite A_1, A_2 in $\mathcal{M}$, A_1 and A_2 are independent over the intersection of their algebraic closures.*
3. *For all finite A_1, A_2 in $\mathcal{M}$, there is a finite C independent from A_1, A_2 such that A_1, A_2 are independent over the intersection of the algebraic closures of $A_1 \cup C$ and $A_2 \cup C$.*
4. *The lattice of algebraically closed subset of $\mathcal{M}^{eq}$ is a modular lattice.*

Proof. The equivalence of (1) and (2) is clear and the equivalence of (1) and (4) is the previous lemma, so we concern ourselves with the implication "(3) implies (2)." We actually show that each instance of (3) implies the corresponding instance of (2).

Let A_1, A_2 be the algebraic closures of two finite subsets of $\mathcal{M}^{eq}$. We must work with sets generated by subsets of $\mathcal{M}$ rather than $\mathcal{M}^{eq}$, so take A_1^*, A_2^* finite subsets of $\mathcal{M}$ such that $A_i \subseteq acl A_i^*$ and, in addition,

(3.1) $A_1^* \perp A_2$ over A_1

(3.2) $A_2^* \perp A_1^*$ over A_2

This ensures $acl(A_1^*) \cap acl(A_2^*) = acl(A_1) \cap acl(A_2)$ by applying first (3.2) and then (3.1). Accordingly, the problem is reduced to the following:

$$A_1^* \perp A_2^* \text{ over } acl(A_1^*) \cap acl(A_2^*).$$

By (3), we have a finite set C independent from $A_1^* A_2^*$ for which

$$A_1^* \perp A_2^* \text{ over } acl(A_1^* \cup C) \cap acl(A_2^* \cup C).$$

Let $A = acl(A_1^* \cup C) \cap acl(A_2^* \cup C)$ and take A_3^* conjugate to A_1^* over $acl(A_2^* \cup C)$, and independent from A_1^* over $A_2^* C$. Then A_3^* is independent from $A_1^* A_2^*$ over A since

$$
\begin{aligned}
rk(A_3^*/A_1^* A_2^* A) &\leq rk(A_3^*/A_1^* A_2^* C) &= rk(A_3^*/A_2^* C) \\
&= rk(A_3^*/A_2^* A) &= rk(A_3^*/A)
\end{aligned}
$$

As A_3^* is independent from $A_1^* A_2^*$ over A and A_1^*, A_2^* are independent over A, A_3^*, A_1^*, A_2^* is an independent triple over A. As A_1^* and A_3^* are conjugate over $acl(A_2^* C)$, they are conjugate over A, and thus $A \subseteq acl(A_3^* C)$. Thus $C \subseteq A \subseteq acl(A_i^* C)$ for all i. For any permutation i, j, k of $1, 2, 3$, we have: $A_i^* \perp A_j^*$ over $A A_k^*$, hence $A_i^* \perp A_j^*$ over $C A_k^*$, and thus $A_i^* \perp A_j^*$ over A_k^*. By Proposition 5.4.1 the triple A_1^*, A_2^*, A_3^* is independent over the intersection of their algebraic closures, and in particular A_1^*, A_2^* are independent over the intersection of their algebraic closures. ∎

Proposition 5.5.4 (Fundamental Rank Inequality, cf. [CHL])

Let $\mathcal{M}$ be $\aleph_0$-categorical, of finite rank, modular, and with the type amalgamation property (cf. §5.1). Let D, D' be 0-definable sets with D' parametrizing a family of definable subsets D_b of D of constant rank r for $b \in D'$. Suppose that E is a 0-definable equivalence relation on D' such that for inequivalent $b, b' \in D'$ we have

$$
rk(D_b) \cap rk(D_{b'}) < r.
$$

Then $rk(D'/E) + r \leq rk\, D$.

Proof. We may assume that both D and D' each realize a unique type over the empty set. Take $b \in D'$ and $a \in D_b$ with $rk(a/b) = r$. Let $C = acl(a) \cap acl(b)$. Thus $a \perp b$ over C by modularity, and $rk(a/C) = rk(a/b) = r$. We will show

$$(*) \qquad\qquad\qquad b/E \in C.$$

Thus $rk(D'/E) \leq rk\, C = rk(aC) - rk(a/C) = rk(a) - r$ as claimed. So we turn to $(*)$.

Let b'/E be a conjugate of b/E over C distinct from b/E, with b' independent from b over C. We seek an element b'' of D' satisfying

$$
tp(b''b/C) = tp(b'b/C); tp(b'', a/C) = tp(b, a/C)
$$

with a, b, b'' independent over C. This amounts to an amalgamation problem for three compatible 2-types: $tp(ba/C)$, $tp(b'b/C)$, $tp(ba/C)$. By the type amalgamation property, this can be done.

In particular, $a \in D_b \cap D_{b''}$ and thus $rk(a/bb'') < r$; but $rk(a/bb'') = rk(a/C) = r$, a contradiction. Thus there is no such conjugate b' and $b \in dcl(C) = C$. ∎

Corollary 5.5.5. *With the hypotheses above, M interprets no Lachlan pseu-doplane.*

Remark 5.5.6. This refers to a combinatorial geometry $(P, L; I)$ of points and lines such that each point is incident with infinitely many lines, two points are incident with only finitely many lines, and dually. The relevance of these structures to the behavior of $\aleph_0$-categorical stable structures was shown in [LaPP], and the corollary settles a question raised in [KLM].

Proof. If $(P, L; I)$ is such a pseudoplane, then after dualizing if necessary we may take $n = rk(L) \geq rk\, P$. We apply the fundamental rank inequality with $D = P$, D_l is the set of points incident with the line l as l varies over a subset D' of L of rank n on which $r = rk\, D_l$ is constant, with E the equality relation. By the axioms for pseudoplanes, the previous proposition applies and yields $rk\, D' + r \leq rk\, P \leq rk\, L = rk\, D'$ and thus $r = 0$, a contradiction. ∎

We give a more precise version of the fundamental rank inequality.

Proposition 5.5.7. *Let D, D' be the loci of single types over the empty set, and D_b a uniformly b-definable family of rank r subsets of D parametrized by D'. Then there is a finite cover $^- : D'' \to D'$ and an equivalence relation E on D'' such that*

1. *$rk(D''/E) = rk\, D - r$;*
2. *For b, b' E-equivalent in D'', we have $rk(D_{\bar{b}} \cap D_{\bar{b}'}) = r$.*

Proof. We work with a, b, c as in the proof of Proposition 5.5.4, but with c finite rather than algebraically closed: so we require $c \in acl(a) \cap acl(b)$ finite, $a \perp b$ over c. Let D'' be the locus of bc over the empty set, with $\overline{b_1 c_1} = b_1$, and with $E(b_1 c_1, b_2 c_2)$ if and only if $c_1 = c_2$ and the types of b_1 over $acl(c_1)$ and of b_2 over $acl(c_2)$ coincide. Then the amalgamation argument yields (2), so $rk(D'')/E = rk(c) = rk(a) - rk(a/c) = rk(D) - rk(a/b) = rk\, D - r$. ∎

5.6 LOCAL CHARACTERIZATION OF MODULARITY

We show in this section that Lie coordinatized structures are modular by reducing the global property of modularity to local properties of the coordinatizing structures.

Definition 5.6.1. *Let M be a structure.*

1. A definable subset D of M is modular if for every finite subset A of M, the structure with universe D and relations the A-definable relations of M restricted to D, is modular.

2. Let $\mathcal{F}$ be a collection of definable subsets of M. Then M is eventually coordinatized by $\mathcal{F}$ if for any $a \in M$ and finite $B \subseteq M$, with $a \notin acl(B)$, *there is $B' \supseteq B$ independent from a over B and a B'-definable member D of $\mathcal{F}$ for which $D \cap acl(aB')$ contains an element not algebraic over B'.*

Lemma 5.6.2. *If $\mathcal{M}$ is eventually coordinatized by a family of modular definable sets, then it is eventually coordinatized by a family of modular definable sets of rank 1.*

Proof. Replace each modular definable set by its definable subsets of rank 1. If $a \in M$ and B is a finite set, take $B' \supseteq B$ independent from a over B and take D definable and modular such that $D \cap acl(aB')$ contains an element b not algebraic over B'.

Take $B_1 \supseteq B'$ such that $rk(b/B_1) = 1$. We may suppose that B_1 is independent from a over B. Let $B_2 = acl(bB') \cap acl(B_1)$. Then $B' \subseteq B_2$, B_2 is independent from a over B, and by modularity of D, b is independent from B_1 over B_2, so $rk(b/B_2) = 1$. Let b' be finite, with $B' \subseteq b' \subseteq B_2$, such that $rk(b/b') = 1$, and let D'_b be the locus of $tp(b/b')$. Then $D'_b \subseteq D$ is rank 1, and is modular since D is. Furthermore, $b \in D_{b'} \cap acl(ab')\setminus acl(b')$, and b' is independent from a over B. ∎

Proposition 5.6.3. *Let $\mathcal{M}$ be $\aleph_0$-categorical of finite rank. If $\mathcal{M}$ is eventually coordinatized by modular definable sets, then $\mathcal{M}$ is modular.*

Proof. By the preceding lemma we may take the coordinatization to be in terms of rank 1 modular sets.

Suppose $\mathcal{M}$ is not modular. Then there are elements a, b and a set E such that $acl(a, E) \cap acl(b, E) = E$, with a and b dependent over E. Take a, b, E with $rk(a/E) + rk(b/E)$ minimal. Then as noted in the proof of Proposition 5.5.3, for any $E' \supseteq E$, independent from a, b over E, a and b remain dependent over $acl(a, E') \cap acl(b, E')$. Thus after applying the eventual coordinatization we may assume in addition that $acl(a, E)$ and $acl(b, E)$ contain elements a_1, b_1 of rank 1 over E, lying in rank 1 modular definable sets D_1, D_2 respectively, defined over E. For the argument below some further expansion of E may be necessary. Specifically, we will assume that E satisfies the following condition:

1. If it is possible to expand E to E' independent from ab over E so that $acl(a, E')$ contains an element a_2 of rank 1 over E' independent from a_1 over E, then the same occurs already over the base E; and similarly for b.

We will also want to assume the following condition for a finite number of elements $a' \in acl(a)$ of rank 1 over E, to be determined below:

2. If there exists E' as described in (1) and $a'' \in D_1$ with $acl(a', E') = acl(a'', E')$, then there is $a^* \in D_1$ for which $acl(a', E) = acl(a^*, E)$; and similarly for b.

After these preliminaries, we may add constants and take $E = acl(\emptyset)$. We will write $0 = acl(\emptyset) = E$. We will show that $a \subseteq acl(a_1 b)$ and $b \subseteq acl(b_1 a)$.

We have $acl(a) \cap acl(b) = 0$, and a, b are dependent. Furthermore, $a_1 \in acl(a)$ has rank 1 and $acl(a_1) \cap acl(b) = 0$, so a_1 and b are independent. As $rk(a/a_1) < rk\, a$, by minimality we have a and b independent over $A = acl(a) \cap acl(a_1, b)$. Since a and b are not independent, A and b are not independent. But $A \subseteq acl(a)$ and hence by minimality of total rank (applied to a finite subset of A, and b) we get $rk(A) = rk\, a$, so $a \subseteq A$. Thus $a \subseteq acl(a_1 b)$; similarly $b \subseteq acl(b_1, a)$.

Now we claim there is a_2 with

$$a_2 \in acl(a); \quad rk(a_2) = 1; \quad a_2 \perp a_1$$

Take b', b_1' conjugates of b, b_1 over a, and independent from b, b_1 over a. Thus $a \subseteq acl(a_1 b')$, and b_1' is independent from a, b. As b depends on a and b_1 does not, we have $rk\, b > rk\, b_1$ and hence we may choose E' containing b_1', independent from a, b, b' over b_1', and some $b_2' \in acl(b', E')$, so that $rk(b_2'/E') = 1$. Now E' is independent from a, b' and $b_2' \in acl(b', E') \subseteq acl\, acl(a, b_1', E') = acl(a, E')$, with a_1 independent from b_2' over E', so the same holds for some conjugate of E' independent from a, b, and then by condition (1) the same holds over 0 for some a_2 in place of b_2'.

Now $a_2 \in acl(a_1 b)$ and thus $a_1 a_2$ depends on b, but $a_1 a_2 \in acl(a)$, so by minimality $a = acl(a_1 a_2)$. Similarly, we get $b = acl(b_1 b_2)$ with b_2 of rank 1. Here no $a_i \in acl(b)$ and no $b_i \in acl(a)$, but any one of a_1, a_2, b_1, b_2 is algebraic over the remainder, and $a_1 \in D_1$. Consider the base set $F = \{a_2, b_2\}$. Then F is independent from b_1 and D_1 contains an element $x = a_1$ such that $acl(x, F) = acl(b_1, F)$. Taking a conjugate E' of F over b_1 free from a, b, (2) applies and yields an element of D_1 that may replace b_1. In the same fashion we may assume $b_2 \in D_1$, and then after reversing the argument, that $a_2 \in D_1$. Then the pair $(a_1 a_2, b_1 b_2)$ violates modularity in D_1. ∎

Corollary 5.6.4. *If $\mathcal{M}$ is Lie coordinatized then $\mathcal{M}$ is modular.*

Proof. The embedded linear and projective geometries are seen to be modular using the last criterion in Proposition 5.5.3, as arbitrary parameters from $\mathcal{M}$ may be replaced by parameters in the geometry. Thus it suffices to show that these geometries eventually coordinatize $\mathcal{M}$.

Let $a \in M$, B a finite subset of M, and $a \notin acl(B)$. One may find $c \in acl(a, B) - acl(B)$ lying in a B-definable coordinatizing projective or affine geometry J. If the geometry is affine, then expand B to $B' = B \cup \{c_0\}$, adding a generic point of J, and replace c by $c - c_0$ in the corresponding linear geometry.

Thus the previous proposition applies. ∎

Definition 5.6.5. *Let a and b be elements of a structure of finite rank. Then b is said to be* filtered *over a if there is a sequence* $\mathbf{b} = b_1, \ldots, b_n$ *with* $rk(b_i/ab_1 \ldots b_{i-1}) = 1$ *and* $acl(ab) = acl(ab)$.

The following was essentially invoked above, and will be applied again subsequently.

Lemma 5.6.6. *Let $\mathcal{M}$ be $\aleph_0$-categorical of finite rank and modular. Then for any a, b in $\mathcal{M}'$, b is filtered over a in $\mathcal{M}'^{\text{eq}}$.*

Proof. Adding constants we may work over the empty set in place of a. We use induction on $n = rk(b)$ and we may suppose $n \geq 1$. We take $b' \in \mathcal{M}'^{\text{eq}}$ with $rk(b/b') = 1$. In particular, b is filtered over b' by b itself, and hence by the previous lemma is independent from b' over $B = acl(b) \cap acl(b')$. Thus $rk(b/B) = rk(b/b') = 1$ and $rk(B) = n - 1$, so by induction after replacing B by a finite set b'' we have a filtration for b' to which we may append b. $\blacksquare$

5.7 REDUCTS OF MODULAR STRUCTURES

In this section we prove the following theorem on reducts of modular structures:

Proposition 5.7.1. *Let $\mathcal{M}$ be $\aleph_0$-categorical of finite rank, and modular. Then every structure $\mathcal{M}'$ interpretable in $\mathcal{M}$ inherits these properties.*

As we will to some extent have both $\mathcal{M}$ and $\mathcal{M}'$ in view throughout the analysis, we adopt the convention that when not otherwise specified, model theoretic notions like rank and algebraic closure that depend on the ambient model will be taken to refer to $\mathcal{M}'$. In any case $\mathcal{M}'$ inherits the $\aleph_0$-categoricity and finite rank. The latter point would however be dubious in general for other notions of rank such as S_1-rank. Furthermore, we cannot assume that the notions of independence in $\mathcal{M}$ and $\mathcal{M}'$ stand in any close relationship.

The main case is that of reducts. In fact, as we can add some parameters and work in $\mathcal{M}^{\text{eq}}$, we may suppose that $\mathcal{M}'$ has as its universe a 0-definable subset of $\mathcal{M}$, and that the structure present on $\mathcal{M}'$ is a reduct of the full structure induced from $\mathcal{M}$. We will refer to this situation as a reduct in (not "of") $\mathcal{M}$.

Lemma 5.7.2. *Let $\mathcal{M}$ be $\aleph_0$-categorical, $\mathcal{M}'$ a reduct, and $\mathbf{a}$ a finite sequence which is algebraically independent in the naive sense: none of its entries is algebraic in $\mathcal{M}'$ over the remainder. Then there is a realization $\mathbf{b}$ of the type of $\mathbf{a}$ in $\mathcal{M}'$, which is algebraically independent in $\mathcal{M}$.*

Proof. Let $\mathbf{b}$ be a realization of the specified type with $acl_{\mathcal{M}}(\mathbf{b})$ as large as possible. If $\mathbf{b}$ contains an entry b which is algebraic over the remainder in $\mathcal{M}$, $\mathbf{b}'$, note that in $\mathcal{M}'$ $b \notin acl(\mathbf{b}')$ and hence there is another realization of the type consisting of $\mathbf{b}'$ extended by some $c \notin acl_{\mathcal{M}}(\mathbf{b}')$. But then $|acl_{\mathcal{M}}(\mathbf{b})| = |acl_{\mathcal{M}}(\mathbf{b}'b)| < |acl_{\mathcal{M}}(\mathbf{b}'c)|$, a contradiction. ∎

Lemma 5.7.3. *Let $\mathcal{M}$ be $\aleph_0$-categorical of finite rank and modular, $\mathcal{M}'$ a reduct in $\mathcal{M}$, and a, b elements of $\mathcal{M}'$ with $rk(b/a) = 1$. Then a is independent from b over $acl(a) \cap acl(b)$.*

We emphasize that our convention applies here, to the effect that the notions used are those of $\mathcal{M}'$ rather than $\mathcal{M}$.

Proof. We will proceed by induction on the rank of a. We may suppose that a and b are algebraically independent, since if $a \in acl(b)$ our claim becomes trivial. By the preceding lemma we may even suppose that they are algebraically independent in $\mathcal{M}$.

Now in $\mathcal{M}$ let $I = (c_1, c_2, \ldots)$ be an infinite $\mathcal{M}$-independent and $\mathcal{M}$-indiscernible sequence over a, with $tp_{\mathcal{M}}(c_i/a) = tp_{\mathcal{M}}(b/a)$. We claim the sequence I is $\mathcal{M}'$-independent over a. For example, $rk(c_{n+1}/ac_1, \ldots, c_n) = 1$ since $rk(c_{n+1}/a) = 1$ and c_{n+1} is not algebraic over $ac_1, \ldots, c_n$ in $\mathcal{M}$, hence certainly not in $\mathcal{M}'$.

The quantity $rk(a/c_1 \ldots c_i)$ as a function of i is eventually constant, say from $i = m$ onward. Let $d = (c_1, \ldots, c_m)$ and $d' = (c_{m+1}, \ldots, c_{2m})$. $rk(a/d) = rk(a/d') = rk(a/dd')$, the latter equality by the choice of m. Thus in $\mathcal{M}'$ we have $a \perp d$ over d', $a \perp d'$ over d, and also $d \perp d'$ over a as checked above. By Proposition 5.4.1, which is applicable to $\mathcal{M}'$, the triple a, d, d' is independent over $A = acl(a) \cap acl(d) \cap acl(d')$. In particular a, c_1 are independent over A.

We now apply the modularity of $\mathcal{M}$. Let $A^* = acl_{\mathcal{M}}(a) \cap acl_{\mathcal{M}}(c_1)$. Since $a \notin acl_{\mathcal{M}}(b)$, also $a \notin acl_{\mathcal{M}}(c_1)$ and thus $a \notin A^*$. By modularity $a \perp_{\mathcal{M}} c_1$ over A^* and by indiscernibility $a \perp_{\mathcal{M}} c_k$ over A^*. As $ac_1 \ldots c_{i-1}$ is $\mathcal{M}$-independent from c_i over a, we find that $a, c_1, c_2, \ldots$ are $\mathcal{M}$-independent over A^*. Hence $a \notin acl_{\mathcal{M}}(c_1, c_2, \ldots)$ and in $\mathcal{M}'$ we have $a \notin acl(d)$, $a \notin A$, and $rk(A) < rk(a)$. Thus by induction $A \perp c_1$ over $A' = A \cap acl(c_1)$, and hence $A \perp c_1$ over A'. Since $tp(ac_1) = tp(ab)$ we have a, b independent over $acl(a) \cap acl(b)$. ∎

Lemma 5.7.4. *Let $\mathcal{M}$ be $\aleph_0$-categorical of finite rank, and modular, and let $\mathcal{M}'$ be a reduct in $\mathcal{M}$. Then every rank 1 subset D of $\mathcal{M}'$ is modular.*

Proof. After absorbing an arbitrary finite set of parameters into the language our claim is that if $\mathbf{a}$, $\mathbf{b}$ are two algebraically independent sequences in D with $acl(\mathbf{a}) \cap acl(\mathbf{b}) = acl(\emptyset)$ in $\mathcal{M}'^{eq}$, then $\mathbf{a}$ and $\mathbf{b}$ are independent. This claim reduces inductively (after further absorption of parameters) to the case in which $\mathbf{a}$ and $\mathbf{b}$ have length 2. In this case if they are not independent, we have $rk(\mathbf{b}/\mathbf{a}) = 1$, and this case was handled in the previous lemma. ∎

Proof of Proposition 5.7.1. It suffices to show that $\mathcal{M}'$ is eventually coordinatized by its rank 1 subsets, since these are modular; we then apply Proposition 5.6.3.

So take $a \notin acl(B)$ with B finite. Let $n = rk(a/B)$. We may find a', c with $a' \in acl(aBc) - acl(Bc)$ and $rk(a/a'Bc) = n - 1$ (cf. Lemma 2.2.3). As $rk(aa'/Bc) = rk(a/Bc)$ this yields

$$rk(a/Bc) = (n - 1) + rk(a'/Bc) \geq rk(a/B)$$

and thus a and c are independent over B and a' has rank 1 over Bc. This shows that $\mathcal{M}'$ is eventually coordinatized by rank 1 subsets. ∎

6

Definable Groups

We study groups definable in Lie coordinatized structures. We will eventually characterize the groups interpretable in Lie coordinatized structures in terms of their intrinsic model theoretic properties. For the stable theory, see the monograph by Poizat [PoGS] and the relevant sections of [Bu, PiGS].

6.1 GENERATION AND STABILIZERS

We work with $\mathcal{M}^{eq}$, and we will consider certain subsets that may meet infinitely many sorts of $\mathcal{M}^{eq}$. In such cases we adopt the following terminology, reflecting the greater generality of this situation relative to the usual context of model theory.

Definition 6.1.1. *Let $\mathcal{M}$ be a many-sorted structure. A subset S of $\mathcal{M}$ is locally definable if its restriction to any sort (equivalently, any finite set of sorts) is definable. In particular, a group is said to be locally definable in $\mathcal{M}$ if its underlying set and its operations are locally definable. When the sorts of $\mathcal{M}$ all have finite rank, a locally definable subset is said to have finite rank if its restrictions to each sort have bounded rank; in this case, the maximum such rank is called the rank of S.*

Remark 6.1.2. One sort of pathology should be noted here. Suppose that in $\mathcal{M}$, the set $dcl(\emptyset)$ meets infinitely many sorts. Let C be a subset of $dcl(\emptyset)$ meeting each sort in a finite set. Then any group structure whatsoever on C is locally definable.

As in §5.5 we say that a structure has the type amalgamation property if Proposition 5.1.15 applies.

We have to be unusually careful with our notation for types in the presence of a group operation, distinguishing $tp(ab)$ (i.e., $tp(a \cdot b)$) from $tp(a, b)$; indeed, the two notions will occur in close proximity.

Lemma 6.1.3. *Let $\mathcal{M}$ be $\aleph_0$-categorical of finite rank. Let G be a locally definable group in $\mathcal{M}^{eq}$, and S a definable subset closed under inversion and generic multiplication: for a, b in S independent, $ab \in S$. Then $H = S \cdot S$ is the subgroup of G generated by S and $rk(H - S) < rk\, S$.*

Proof. We show first that the product of any three elements a_1, a_2, a_3 of S lies in $S \cdot S$; this shows both that $S \cdot S$ is a subgroup of G, and that $S \subseteq S \cdot S$ (take $a_2 = a_1^{-1}$).

Given a_1, a_2, a_3 we take $u \in S$ independent from a_1, a_2, a_3 and of maximal rank. Let $b_1 = a_1 u$ and $b_2 = u^{-1} a_2$. Then $b_1, b_2 \in S$. Furthermore, b_2, a_3 are independent and thus $b_2 a_3 \in S$. But $a_1 a_2 a_3 = b_1 \cdot b_2 a_3$.

It remains to consider $rk(H - S)$. Let $a_1 a_2 \in H$ have rank at least $rk(S)$. Take $u \in S$ of maximal rank independent from a_1, a_2. Then $a_2 u$ belongs to S and is independent from a_1. Thus $b = a_1 a_2 u$ is also in S and

$$rk(b/u) = rk(a_1 a_2/u) = rk(a_1 a_2) \geq rk\, S$$

and thus equality holds, and b and u are independent. We therefore have $a_1 a_2 = bu^{-1} \in S$. Thus $rk(H - S) < rk\, S$. ∎

Lemma 6.1.4. *Let $\mathcal{M}$ be $\aleph_0$-categorical of finite rank, with the type amalgamation property. Let G be a locally 0-definable group of finite rank k in $\mathcal{M}^{eq}$, and $S \subseteq G$ the locus of a complete type over $acl(\emptyset)$, of rank k. Then $S \cdot S^{-1}$ generates a definable subgroup of G.*

We do not claim that S itself generates a definable subgroup; for example, if S reduces to a single element then the group in question is trivial. On the other hand, the statement of the lemma is equivalent to the claim that S generates a coset of a definable group under the affine group operation $ab^{-1}c$.

Proof. Let $X = \{ab^{-1} : a, b \in S; rk(a, b) = 2k\}$. Note that for $a, b \in S$ independent of rank k, $rk(a, ab^{-1}) = rk(a, b) = 2k$ and thus also a, ab^{-1} are independent of rank k. We claim that the previous lemma applies to X, and that the groups generated by $S \cdot S^{-1}$ and by X coincide. In any case X is closed under inversion. We show now that X is closed generically under the operation ab^{-1}, and hence also under multiplication.

Let $c_1, c_2 \in X$ be independent, $c_i = a_i b_i^{-1}$ with $a_i, b_i \in S$, $rk(a_i, b_i) = 2k$. We may suppose that (a_1, b_1) is independent from (a_2, b_2) and hence that a_1, a_2, b_1, b_2 is an independent quadruple. We seek d independent from this quadruple satisfying

$$tp(d/c_1) = tp(b_1/c_1); \quad tp(d/c_2) = tp(b_2/c_2).$$

As S is a complete type over $acl(\emptyset)$ and b_i is independent from c_i, this is a type amalgamation problem of the sort that can be solved. The type of d now ensures the solvability of the equations

$$c_1 = a_1' d^{-1}; \quad c_2 = a_2' d^{-1}$$

with a_1', a_2' in S. Thus $c_1 c_2^{-1} = a_1' a_2'^{-1}$. We claim that this forces $c_1 c_2^{-1}$ into S, with a_1', a_2' as witnesses. Since $a_i' \in dcl(a_i, b_i, d)$, we have a_1' and a_2'

independent over d. Also $rk(a_i', b_i, d) = rk(a_i, b_i, d) = 3k$, so a_i and e are independent. Thus a_1' and a_2' are independent. Thus $c_1 c_2^{-1} \in X$.

Now suppose $a, b \in S$. Take $d \in S$ independent from a. Then $ab^{-1} = (ad) \cdot (bd)^{-1} \in X \cdot X$. Thus $S \cdot S^{-1}$ and X generate the same subgroup. ∎

Lemma 6.1.5. *Let M be $\aleph_0$-categorical of finite rank. Let G be a locally definable group in M^{eq}, and S a definable subset generically closed under the ternary operation $ab^{-1}c$ (an affine group law). Then S lies in a coset C of a definable subgroup H of G, with $rk(C - S) < rk\,S$.*

Proof. We consider $X = \{ab^{-1} : a, b \in S \text{ independent}\}$. The condition on S implies that X is generically closed under multiplication and Lemma 6.1.3 applies, so X generates a definable subgroup H with $rk(H - X) < rk\,X$.

We claim that S lies in a single coset C of H. Indeed, if $a, b \in S$ and $c \in S$ is independent from a, b, then $ab^{-1} = (ab^{-1}c)c^{-1} \in H$.

Lastly, we claim that $rk(C - S) < rk\,S$. Let $a_o \in S$ and let $h \in H$ be independent from a, of maximal rank. Then $h \in X$ and $h = ab^{-1}$ with a, b, a_o independent. Then $ha_o = ab^{-1}a_o \in S$. Thus Ha_o lies in S up to a set of smaller rank. ∎

Definition 6.1.6. *Let $h : G_1 \to G_2$ be a map between groups. Then h is an affine homomorphism if it respects the operation $ab^{-1}c$.*

Lemma 6.1.7. *Let M be $\aleph_0$-categorical of finite rank. Let G, H be locally 0-definable groups in M^{eq}, S a 0-definable subset of G, and $h : S \to H$ a 0-definable function.*

1. *If S is generically closed under the affine group operation $ab^{-1}c$ and h generically respects this operation, then h extends to an affine group homomorphism with domain the coset of a definable subgroup generated by S (under the affine group operation).*
2. *If S is generically closed under the operation ab^{-1} and h generically respects this operation, then h extends to a group homomorphism defined on the subgroup of G generated by S.*

Proof. Consider the graph Γ of the map h as a definable subset of the product group $G \times H$. Then Γ satisfies the hypotheses of Lemma 6.1.5 or Lemma 6.1.3, respectively. Thus in case (1) under the affine group operation Γ lies in a coset $\bar{\Gamma}$ of a definable subgroup of $G \times H$, with $rk(\bar{\Gamma} - \Gamma) < rk\,\Gamma$, and in case (2) Γ lies in a definable subgroup $\bar{\Gamma}$ of $G \times H$, with $rk(\bar{\Gamma} - \Gamma) < rk\,\Gamma$. Here $\bar{\Gamma}$ will again be the graph of a function, as otherwise $\bar{\Gamma}$ will contain a translate of Γ disjoint from Γ, violating the rank condition. $\bar{\Gamma}$ is of course the graph of the desired extension of Γ in either case. ∎

In the next lemma the avoidance (or neutralization) of the pathological case referred to at the outset is particularly important.

Lemma 6.1.8. *Let $\mathcal{M}$ be $\aleph_0$-categorical of finite rank, with the type amalgamation property. Let G be a locally definable group in $\mathcal{M}^{eq}$ of bounded rank which is abelian, of bounded exponent. Then for any definable subset S of G, the subgroup generated by S is definable.*

Proof. We may take everything (locally) 0-definable. We may also suppose that S generates G. Our statement then amounts to the claim that G meets only finitely many sorts of $\mathcal{M}^{eq}$. The case of rank 0 will play a key role below; in this case we are considering a finitely generated subgroup of a locally finite group, so the group in question is finite and hence definable.

Now let $k = rk\, G$. Replacing S by a larger set if necessary we may suppose S has rank k. Let $S_\circ$ be the locus of some type of rank k over $acl(\emptyset)$ contained in S. Then under the affine group operation $S_\circ$ generates a coset of a definable subgroup H of rank k (Lemma 6.1.4). Now work in G/H. $S+H$ meets a finite number of sorts, and $k \geq rk(S+H) \geq rk(S/H) + rk(H) \geq rk(S/H) + k$, so S/H is finite and therefore generates a finite subgroup of the locally finite quotient G/H, as noted at the outset. $\blacksquare$

We now turn to the notion of the stabilizer of a definable set S. Though it is tempting to define this as the group of $g \in G$ such that gS and S agree modulo sets of smaller rank, this tends to define the trivial subgroup and is therefore not useful. Note that most of our underlying geometries do not in any sense have "Morley degree" 1, or even finite degree.

Definition 6.1.9. *Let $\mathcal{M}$ have finite rank, let G be a definable group in $\mathcal{M}$, and let D, D' be complete types over $acl(\emptyset)$, contained in G, with $rk\, D = rk\, D' = r$. Then*
1. $Stab_\circ(D, D') = \{g \in G : rk(Dg \cap D') = r\}$.
2. $Stab_\circ(D) = Stab_\circ(D, D)$ *and* $Stab(D)$ *(the full stabilizer of D) is the subgroup of G generated by $Stab_\circ(D)$.*

Though we claim that $Stab_\circ(D)$ is generically closed under multiplication, it will not in general actually be a subgroup.

Example 6.1.10. *Let (V, Q) be an infinite dimensional orthogonal space over a finite field of characteristic 2, with the associated symplectic form degenerate, with a 1-dimensional radical K on which Q is nonzero. Let $D = \{x \neq 0 : Q(x) = 0\}$. Then $Stab_\circ(D) = V - (K - (0))$ is not a subgroup.*

Lemma 6.1.11. *Let $\mathcal{M}$ be $\aleph_0$-categorical of finite rank with the type amalgamation property, G a 0-definable group in $\mathcal{M}^{eq}$. Let D, D', and D'' be complete types over $acl(\emptyset)$ of rank r contained in G. If $a \in Stab_\circ(D, D')$ and $b \in Stab_\circ(D', D'')$ are independent, then $ab \in Stab_\circ(D, D'')$.*

Proof. $rk(Da) \cap D' = r = rk(D''b^{-1} \cap D')$, so by the corollary to Proposition 5.1.15 we have also $rk(Da \cap D' \cap D''b^{-1}) = r$, and after multiplication on the right by b we have $rk(Dab \cap D'') = r$. ∎

Lemma 6.1.12. *Let $\mathcal{M}$ be $\aleph_0$-categorical of finite rank with the type amalgamation property, G a 0-definable group in $\mathcal{M}^{eq}$, and D a complete type over $acl(\emptyset)$. Then*

$$Stab(D) = Stab_o(D)\, Stab_o(D);$$
$$rk(Stab(D) - Stab_o(D)) < rk(Stab_o(D)).$$

Proof. Lemmas 6.1.3 and 6.1.11. ∎

Lemma 6.1.13. *Let $\mathcal{M}$ be $\aleph_0$-categorical of finite rank with the type amalgamation property, G a 0-definable group in $\mathcal{M}^{eq}$, and D a complete type over $acl(\emptyset)$ with $rk\, D = rk\, G$. Then $[G : Stab(D)] < \infty$.*

Proof. It suffices to show that $rk\, Stab_o(D) = rk\, G$. Let a, b be independent elements of D of rank $r = rk\, G$ and $c = a^{-1}b$. Then $rk(b, c) = 2r$ so $rk(b/c) = r$, and $b \in D \cap Dc$. Thus $c \in Stab_o D$. As c has rank r, we are done. ∎

Lemma 6.1.14. *Let $\mathcal{M}$ be $\aleph_0$-categorical of finite rank with the type amalgamation property, G a 0-definable group in $\mathcal{M}^{eq}$, D a 0-definable subset of G with $rk\, D = rk\, G$, and suppose that G has no proper 0-definable subgroup of finite index. Then there are pairwise independent $a_1, a_2, a_3 \in D$ with $a_1 a_2 = a_3$.*

Proof. We may take D to be the locus of a complete type over $acl(\emptyset)$. Then by the preceding lemma and our hypothesis $Stab(D) = G$. By Lemma 7 $rk(D \cap Stab_o\, D) = rk\, G$. Pick $a_1 \in D \cap Stab_o\, D$ of rank $rk\, G$ and $a_3 \in Da_1 \cap D$ with $rk(a_3/a_1) = rk\, G$. Then set $a_2 = a_3 a_1^{-1}$. ∎

6.2 MODULAR GROUPS

Definition 6.2.1. *Two subgroups H_1, H_2 of a group G are said to be commensurable if their intersection has finite index in each. This is an equivalence relation. When G has finite rank, this is equivalent to $rk(H_1) = rk(H_2) = rk(H_1 \cap H_2)$.*

Lemma 6.2.2. *Let $\mathcal{M}$ be $\aleph_0$-categorical of finite rank and modular. Let G be a definable group in $\mathcal{M}$, and H_d a subgroup defined uniformly from the parameter d for d varying over a definable set D. Let $E(d, d')$ hold if and only if H_d and $H_{d'}$ are commensurable. Then the relation E has finitely many equivalence classes.*

Proof. Choose $d \in D$ of maximal rank, $a \in G$ of maximal rank over d, and b in $H_d a$ of maximal rank over a, d. Let $B = acl(b) \cap acl(d, a)$. Let d', a' be conjugate to d, a over b and independent from d, a over b. Then b, d, a, and d', a' are independent over B by modularity and the choice of d', a'. Thus $rk(b/aa'dd') = rk(b/B) = rk(b/ad)$ and $rk(H_d a \cap H_{d'} a') = rk(H_d a)$. Therefore $rk(H_d \cap H_{d'}) = rk(H_d)$, in other words $E(d, d')$ holds. Thus $d/E \in B$.

Furthermore, as $(H_d \cap H_{d'})a'a^{-1}$ is nonempty, $a'a^{-1}$ lies in $H_d H_{d'} = X_d X_{d'}(H_d \cap H_{d'})$ for sets $X_d, X_{d'}$ of coset representatives of the intersection in H_d, $H_{d'}$, respectively. Thus $rk(a'/a, d, d') \leq rk H_d$ and hence $rk(a/B) \leq rk H_d$. Now we compute $rk(d/E)$:

$$\begin{aligned}
rk(d, a, b) &= rk(d) + rk(a) + rk(b/a, d) = rk(a) + rk\,G + rk\,H_d \\
&= rk(b) + rk(a/b) + rk(d/a, b) \\
&\leq rk\,G + rk\,H_d + rk(d/(d/E))
\end{aligned}$$

showing $rk(d/(d/E)) = rk(d)$ and $rk((d/E)) = 0$, so $d/E \in acl(\emptyset)$. ∎

The next proposition, for which we give a purely model theoretic argument, can be proved in greater generality as a purely group theoretic statement [Sch, BeLe]. This was drawn to our attention by Frank Wagner, who has generalized the result even further [Wa].

Proposition 6.2.3. *Let $\mathcal{M}$ be $\aleph_0$-categorical of finite rank with the type amalgamation property and modular. Let G be a 0-definable group in $\mathcal{M}$, and H a definable subgroup. Then H is commensurable with a group defined over $acl(\emptyset)$.*

Proof. Let $H = H_d$ have defining parameter $d \in D$, with D a complete type over $acl(\emptyset)$. Let $E(d, d')$ be the equivalence relation: H_d, $H_{d'}$ are commensurable. As this has finitely many classes and D realizes a unique type over $acl(\emptyset)$, all groups H_d ($d \in D$) are commensurable.

Define $B = \{g \in G : \text{For some } d \in D \text{ independent from } g, g \in H_d\}$. By the corollary to Proposition 5.1.15,

$$\text{For } b_1, b_2 \text{ in } B \text{ independent, } b_1 b_2^{-1} \in B.$$

Thus by Lemma 6.1.3, $H = \langle B \rangle$ is a definable subgroup of G with $rk(H - B) < rk\,H$. Let $h \in H$ be an element of maximal rank. Then $h \in B$. Take $d \in D$ independent from h with $h \in H_d$. Then $rk(h) \leq rk\,H_d$ and thus $rk\,H \leq rk\,H_d$. On the other hand any element of H_d independent from d is in B, so $rk(H \cap H_d) \geq rk\,H_d$. This shows that H and H_d are commensurable. ∎

Proposition 6.2.4. *Let $\mathcal{M}$ be $\aleph_0$-categorical of finite rank with the type amalgamation property, and modular. Let G be a 0-definable group in $\mathcal{M}$. Then*

G has a finite normal subgroup N such that G/N contains an abelian subgroup of finite index.

Proof. Let $Z^* = \{g \in G : [G : C(g)] < \infty\}$. We work mainly in $G^2 = G \times G$. For $a \in G$ let H_a be the subgroup $\{(x, x^a) : x \in G\}$ of G^2. Define $E(a, a')$ as follows: H_a and $H_{a'}$ are commensurable. This is an equivalence relation with finitely many classes. Notice that $E(a, a')$ holds if and only if $Z^*a = Z^*a'$: $E(a, a')$ holds if and only if on a subgroup G_1 of G of finite index we have $x^a = x^{a'}$; that is, $G_1 \leq C(a'a^{-1})$, $a'a^{-1} \in Z^*$.

Thus we have proved that Z^* is of finite index in G and we may replace G by Z^*. Then any element of G has finitely many conjugates and thus for $x, y \in G$, the commutator $[x, y]$ is algebraic over x and over y. In particular for $x, y \in G$ independent, the commutator $[x, y]$ is algebraic over $\emptyset$. On the other hand, every commutator $[x, y]$ can be written as $[x, y']$ with y' independent from x, since $C(x)$ has finite index in G. Thus $N = G'$ is finite, and G/N is abelian. ∎

As this result tends to reduce the study of definable groups to the abelian case, we will generally restrict our attention to abelian groups in the sequel, even when this hypothesis is superfluous.

Lemma 6.2.5. *Let $\mathcal{M}$ be $\aleph_0$-categorical of finite rank with the type amalgamation property, and modular. Let A be a 0-definable abelian group in $\mathcal{M}$, and $D \subseteq A$ the locus of a complete type over $\mathrm{acl}(\emptyset)$, S the stabilizer of D in A. Then*

1. *$rk\, S = rk\, D$.*
2. *D is contained in a single coset of S.*
3. *If D' is the locus of another complete type over $\mathrm{acl}(\emptyset)$ of the same rank, and if $\mathrm{Stab}_o(D, D')$ is nonempty, then $\mathrm{Stab}_o(D, D')$ agrees with a coset of S up to sets of smaller rank, and $\mathrm{Stab}(D') = S$.*
4. *If $a, b \in S$ are independent with the same type over $\mathrm{acl}(\emptyset)$, then $a - b \in \mathrm{Stab}_o(D)$.*

Proof.

Ad 1. We apply the fundamental rank inequality of Proposition 5.5.4 taking both 0-definable sets to be G, and $G_a = D + a$, relative to the equivalence relation $E(a, b)$: $a - b \in S$. Then for inequivalent elements a, b the intersection $G_a \cap G_b$ has lower rank, so the fundamental rank inequality 5.5.4 applies and yields

$$rk(A/S) \leq rk\, A - rk\, D, \text{ hence } rk(S) \geq rk(D).$$

The opposite inequality is elementary: if $s \in S$ has maximal rank and $d \in D$ has maximal rank over s, with $d + s \in D$, then $rk(d + s/d) = rk(s/d) = rk\, S$, so $rk(S) \leq rk(D)$. Notice also that $a + S$ meets D in a set of rank $rk\, D$.

Ad 2. We have seen that some coset of S meets D in a set of rank $rk\,D$. There can be only finitely many such cosets, so they lie in $acl(\emptyset)$, and as D realizes a single type over $acl(\emptyset)$, there is only one such coset, and it contains D.

Ad 3. According to Lemma 6.1.11, if $a \in Stab(D)$, $b \in Stab(D, D')$ are independent, then $a + b \in Stab(D, D')$. Thus under the stated hypothesis $Stab(D, D')$ contains most of a coset of S, up to a set of lower rank. Conversely if $a, b \in Stab(D, D')$ are independent then by the same lemma $a - b \in S$, so $Stab(D, D')$ agrees with a single coset of S modulo sets of lower rank. Replacing a by $-a$ we find that $Stab(D', D)$ agrees with a single coset of $Stab(D')$ modulo sets of lower rank and thus S and $Stab(D')$ agree modulo sets of lower rank; as they are groups, they are equal.

Ad 4. By $(1, 2)$ we have $rk\,SD = rk\,D = rk\,Stab_\circ\,D$. Thus the corollary to Proposition 5.1.15 applies. ∎

Lemma 6.2.6. *Let $\mathcal{M}$ be $\aleph_0$-categorical of finite rank with the type amalgamation property, and modular. Let A be a 0-definable group in $\mathcal{M}^{eq}$ of rank n. Then there is a sequence of subgroups $(0) = A_\circ \triangleright A_1 \triangleright \ldots \triangleright A_n = A$ with $rk(A_i/A_{i-1}) = 1$, and all A_i defined over $acl(\emptyset)$.*

Proof. We may replace A by a quotient modulo a finite normal subgroup of a subgroup of a finite index, so we may take A abelian. It suffices then to find a subgroup of rank 1 defined over $acl(\emptyset)$ as we may factor it out and proceed inductively. Let D be the locus in A of a complete rank 1 type over an algebraically closed set. By Lemma 6.2.5 the stabilizer of D in A is a rank 1 subgroup. By Proposition 6.2.3 it is commensurable with a group defined over $acl(\emptyset)$. ∎

Lemma 6.2.7. *Let $\mathcal{M}$ be $\aleph_0$-categorical of finite rank with the type amalgamation property, G, H 0-definable groups in $\mathcal{M}^{eq}$, $D \subseteq G$ the locus of a complete type over $acl(\emptyset)$ with $rk\,D = rk\,G$, $f : D \to H$ 0-definable, and suppose that for any independent triple $a_1, a_2, a_3 \in D$ for which $a_1 a_2^{-1} a_3 \in D$, we have $f(a_1 a_2^{-1} a_3) = f(a_1)f(a_2)^{-1}f(a_3)$. Then f extends to a definable affine homomorphism from the coset in G generated by D, to H.*

Proof. Let $r = rk\,G$. We first define a function $h : Stab(D) \to H$. Let $S^* = \{a \in Stab(D) : rk(a) = rk\,G\}$. Then $S^* \subseteq Stab_\circ(D)$.

If $a \in S^*$ then $a = b_1 b_2^{-1}$ with b_1, b_2 independent elements of rank r in D. We define $h(a) = h(b_1)h(b_2)^{-1}$ and we must check that this is in fact well defined. Suppose also $a = b_3 b_4^{-1}$ with b_3, b_4 independent of rank r in D. Take further b_5, b_6 independent and of rank r, with $a = b_5 b_6^{-1}$, such that $rk(b_5, b_6/ab_1 b_2 b_3 b_4) = r$. Then b_1, b_2, b_6 and b_3, b_4, b_6 are independent triples with $b_6 b_2^{-1} b_1 = b_6 b_4^{-1} b_3 = b_5$, so applying the affine homomorphism law for f and cancelling $f(b_6)$, we get $f(b_1)b(b_2)^{-1} = f(b_3)f(b_5)^{-1}$ and f is well defined on S^*.

In order to extend h from S^* to $Stab(D)$, we show that part (2) of Lemma 6.1.7 applies. Let $a, b \in S^*$ be independent, and $c = ab^{-1}$. Certainly $c \in S^*$. We have $rk(D \cap Da) = rk(D \cap Db) = rk\,D$ and thus by the corollary to Proposition 5.1.15 $rk(D \cap Da \cap Db) = r$. Take $d_1 \in D \cap Da \cap Db$ of rank r over a, b, and set $d_2 = d_1 a^{-1}$, $d_3 = d_1 b^{-1}$. Thus $a = d_2^{-1} d_1$, $b = d_3^{-1} d_1$, $c = d_2^{-1} d_3$, with pairs of independent elements of rank r. The resulting formulas $h(a) = f(d_2)^{-1} f(d_1)$ and so forth combine to give $h(c) = h(a) h(b)^{-1}$, as required. Thus we may now take h to be a homomorphism from $S = Stab(D)$ to H.

D is contained in a single left coset C of S. For $b \in D$ we define a map $f_b : C \to H$ by $f_b(x) = f(b) h(b^{-1} x)$. This is an affine homomorphism from $C \to H$ which agrees with f on elements of D independent from b, using the basic property of f and the definition of h. Our final point will be that f_b is independent of $b \in D$ and therefore gives the desired extension f^* of f to C.

To see that f_b does not depend on b it suffices to prove $f_b = f_{b'}$ for $b, b' \in D$ independent. For any $c \in C$ we have $h(b^{-1} c) h(b'^{-1} c)^{-1} = h(b^{-1} b') = f(b^{-1}) f(b')$ and thus $f_b(c) = f_{b'}(c)$. ∎

Lemma 6.2.8. *Let $\mathcal{M}$ be $\aleph_0$-categorical of finite rank with the type amalgamation property and modular. Let A_1, A_2 be 0-definable abelian groups in $\mathcal{M}^{eq}$. Suppose that any $acl(\emptyset)$-definable subgroup of $A_1 \times A_2$ is 0-definable, and that $acl(\emptyset) \cap A_1 = (0)$. Let C be a finite set with $acl(C \cap A_1) \subseteq C$, and let $a_2 \in A_2$ have maximal rank over C. Then*

1. *$acl(a_2, C) \cap A_1 \subseteq dcl(a_2, acl(C))$;*
2. *If no proper definable subgroup of A_2 of finite index is definable over $acl(\emptyset)$, then $acl(a_2, C) \cap A_1 = [dcl(a_2) \cap A_1] + [C \cap A_1]$.*

Proof. Let $a_1 \in acl(a_2, C) \cap A_1$. Let D be the locus of (a_1, a_2) over $acl(C)$, and $S = Stab(D)$ in $A_1 \times A_2$. By Proposition 6.2.3, the group S is commensurable with a group S' defined over $acl(\emptyset)$; by hypothesis, S' is 0-definable.

By Lemma 6.2.6 D is contained in a coset of S, and hence in a finite union of cosets of $S \cap S'$; as D is the locus of a complete type over an algebraically closed set, D is contained in a single coset X of $S \cap S'$. In particular $S \subseteq D - D \subseteq S'$. Thus X is a coset of S'.

Now $S \cap [A_1 \times (0)]$ is finite, since a_1 is algebraic over a_2, C. Thus $S' \cap [A_1 \times (0)]$ is finite, and since $acl(\emptyset) \cap A_1 = (0)$, we conclude that $S' \cap [A_1 \times (0)] = 0$. Let $\pi_2 : A_1 \times A_2 \to A_2$ be the projection. Thus $S' = S \cap S'$ is the graph of a homomorphism from $\pi_2 S'$ to A_1 and X is the graph of an affine homomorphism $f : \pi_2 X \to A_1$. As X is definable over $acl(C)$, $f(a_2) \in dcl(a_2, acl(C))$. This proves the first claim. Under the hypothesis of (2), S' is the domain of a homomorphism $h : A_2 \to A_1$ and $f - h$ is a constant $a \in A_1$, so $a \in acl(C) \cap A_1 = C \cap A_1$ and $a_1 = h(a_2) + a \in [dcl(a_2) \cap A_1] + [C \cap A_1]$

as claimed. ∎

Remark 6.2.9. With the above hypotheses and notation, the same result can be proved, with the same proof, for the affine space S_1 over A_1, assuming $acl(C) \cap S_1 = dcl(C) \cap S_1$.

Proposition 6.2.10. *Let $\mathcal{M}$ be $\aleph_0$-categorical of finite rank with the type amalgamation property, and modular. Let A be a 0-definable rank 1 abelian group in $\mathcal{M}^{eq}$. Assume that $acl(\emptyset) \cap A = (0)$ and that A has no proper $acl(\emptyset)$-definable subgroup of finite index. Then there is a finite field F such that A has a definable vector space structure over F for which linear dependence coincides with algebraic dependence.*

Proof. Let F be the ring of $acl(\emptyset)$-definable endomorphisms of A. Our assumptions on A imply that F is a division ring and by $\aleph_0$-categoricity of $\mathcal{M}$, F is finite; thus it is a finite field. Take A now as a vector space over F.

We show by induction on n that any n algebraically dependent elements $a_1, \ldots, a_n$ of A will be linearly dependent. For the inductive step, suppose that $a_1, \ldots, a_{n+1}$ are algebraically dependent and $a_1, \ldots, a_n$ are algebraically independent. Thus $a = a_{n+1} \in acl(a_1, \ldots, a_n)$ and we wish to express a as an F-linear combination of $a_1, \ldots, a_n$. Let D be the locus of $(a_1, \ldots, a_n, a)$ over $acl(\emptyset)$, and S its stabilizer. We have $rk\, S = rk\, D = n \times rk\, A = n$, with D contained in a coset of S. Let T be the projection of S onto the first n coordinates. As the projection of D to these coordinates contains $(a_1, \ldots, a_n)$ and is contained in a coset of T, $rk\, T = n$. Therefore the kernel of this projection has rank 0 and is finite, and T has finite index in A^n. By our hypotheses on A, the kernel is trivial and $T = A^n$ (consider the intersection of T with the standard copies of A in A^n). In other words S is the graph of a homomorphism $h : A^n \to A$, i.e., $h = (\alpha_1, \ldots, \alpha_n)$ with $\alpha_i : A \to A$ definable over $acl(\emptyset)$. We claim naturally that $a = \sum \alpha_i a_i$ with $\alpha_i \in F$.

As D is contained in a coset of S, for $(x_1, \ldots, x_n, y)$ and $(x'_1, \ldots, x'_n, y')$ in D, we get $y - y' = h(\mathbf{x} - \mathbf{x}') = \sum_i \alpha_i(x_i - x'_i)$ and thus $y - \sum_i \alpha_i x_i$ is a constant on D, belonging to $acl(\emptyset) \cap A = (0)$. This proves our claim. ∎

Lemma 6.2.11. *Let $\mathcal{M}$ be Lie coordinatized, and A an infinite abelian group interpreted in $\mathcal{M}$ without parameters. Suppose that A has no nontrivial $acl(\emptyset)$-definable proper subgroup, and that $acl(\emptyset) = dcl(\emptyset)$. Then A is part of a basic linear geometry in $\mathcal{M}$.*

Proof. By the previous proposition A has a vector space structure over a finite field F such that algebraic dependence coincides with F-linear dependence. Let P be the corresponding projective space. Then P is nonorthogonal to some $acl(\emptyset)$-definable projective Lie geometry PJ, and there is then a 0-definable bijection between these geometries. Taking a cover of $\mathcal{M}$ if necessary, PJ will be the projectivization of a basic linear geometry J. By Lemma 2.4.7 there is a 0-definable isomorphism of A with J, so A is a basic linear geometry. ∎

6.3 DUALITY

We will be dealing with groups definable in weakly Lie coordinatized structures below. As we make some use of envelopes, we observe that by Lemma
6.2.6 any such group is nonmultidimensional in the sense that it lies in a part
of the structure which is coordinatized by a finite number of Lie geometries,
each defined over $acl(\emptyset)$. (More precisely: first adjust the base language tem-
porarily so that the group in question is viewed as defined over $acl(\emptyset)$.) In
particular, only a finite number of quadratic geometries are involved, and after
naming the Witt classes by introducing finitely many algebraic parameters, we
may work in a Lie coordinatized structure. This being the case, it suffices to
state the results in the Lie coordinatized setting; they then apply in the weakly
Lie coordinatized setting as well.

Definition 6.3.1. *If M is a structure, and A a group of prime exponent p interpreted in M, then A^* denotes the group of M^{eq}-definable homomorphisms
from A to a cyclic group of order p (equivalently the set of definable F-linear
maps from A to the field F of order p).*

Note that the elements of A^* are almost determined by their kernels, which
are definable subgroups of A. However, we do not necessarily have $A^* \subseteq A^{\mathrm{eq}}$;
for example, A may be one side of a polar geometry.

The reader should bear in mind that the abelian groups A of this section are
not intended to be reminiscent of affine geometries.

Proposition 6.3.2. *Let M be a Lie coordinatized structure, A a 0-definable
group in M^{eq} of prime exponent p. Then A^* and the evaluation map $A \times
A^* \to \mathbb{F}_p$ are 0-definable in M^{eq}. If A has no nontrivial proper 0-definable
subgroups, then either $A^* = (0)$ or the pairing $A \times A^* \to \mathbb{F}_p$ is a perfect
pairing (the annihilator of each factor in the other is trivial).*

Proof. A^* is a locally definable group. Arrange the sorts of M^{eq} in some
order and let D_n be the definable subset of A^* consisting of elements which
lie in the first n sorts.

Our first claim is that $rk\, A^*$ is finite, bounded by $rk\, A$. Fix a definable subset
D of A^*, and suppose $rk\, D > rk\, A$. We apply Proposition 5.2.2 concerning
the sizes of envelopes. Accordingly, the number of elements of D is a polynomial of degree $2\, rk\, D$ in the variables used there, and similarly for A. Taking
envelopes of large and constant dimension, we deduce that $D \cap E$ eventually is
larger than $A \cap E$, while (again for large enough envelopes) $D \cap E \subseteq (A \cap E)^*$;
this is a contradiction.

We apply Lemma 6.1.8 and deduce that for any n the subgroup A_n^* generated by D_n is 0-definable. Let K_n be the annihilator in A of A_n^*. The decreasing chain K_n of 0-definable groups must stabilize with $K_n = K$ constant from some point on. We may factor out K and suppose $K = (0)$ (note in passing that the last part of the proposition will be covered by the argument from this point on).

After these preliminaries we see that $A \times A_n^* \to F$ is a perfect pairing for all large n. Therefore, with n, n' fixed, looking at the same situation in large finite envelopes, we find that $A_n^* \cap E = A_{n'}^*$ in such envelopes. Thus A_n^* is independent of n for n large, and $A_n^* = A^*$. ∎

We note that one can form a structure consisting of a set D and a vector space V, with a generic interaction of D with V in which the elements of D act linearly on V. The foregoing proposition will fail for this structure, which is not Lie coordinatizable.

We now mention a variation of somewhat greater generality:

Lemma 6.3.3. *Let M be a Lie coordinatized structure, A a 0-definable group in M^{eq} of finite exponent n, and A^* the definable $\mathbb{Z}/p\mathbb{Z}$-dual of A. Then A^* and the pairing $A \times A^* \to \mathbb{Z}/n\mathbb{Z}$ are interpretable in M. Furthermore, any definable subgroup B of A of finite index is an intersection of the kernels of elements of A^*.*

Proof. The definability of A^* is just as before. For the final statement, since A/B has exponent dividing n, it is perfectly paired with its $\mathbb{Z}/n\mathbb{Z}$ dual. ∎

Lemma 6.3.4. *Let M be a Lie coordinatized structure, A a 0-definable vector space in M^{eq} relative to a finite field K of characteristic p. Let A^* be the definable $\mathbb{Z}/p\mathbb{Z}$-dual of A, and Tr the trace from K to the prime field. Then A^* can also be given a K-space structure, and there is then a definable K-bilinear map $\mu : A \times A^* \to K$ such that $Tr\,\mu(a, f) = f(a)$ for $(a, f) \in A \times A^*$. This pairing makes A^* the full definable K-linear dual of A.*

Proof. Let A' be the space of all definable K-linear maps of A to K. Let $Tr : A' \to A^*$ be defined by $Tr(f)(a) = Tr(f(a))$. If $Tr(f) = 0$ then for $a \in A$ and $\alpha \in K$ we have $Tr(\alpha f(a)) = Tr(f)(\alpha a) = 0$, and thus $f(a) = 0$ by the nondegeneracy of the bilinear form $Tr(xy)$. Thus Tr embeds A' into A^*. Conversely, if $g \in A^*$ then for $a \in A$ the linear map $g_a : K \to F$ defined by $g_a(\alpha) = g(\alpha a)$ must have the form $Tr(\gamma_a \alpha) = g(\alpha a)$ for a unique $\gamma_a \in K$. Letting $f(a) = \gamma_a$ we get $Tr(f) = g$, and f is K-linear since $f(\alpha\beta a) = Tr(\beta\gamma_a\alpha)$. Thus Tr identifies the K-linear dual with the F-linear dual. Let μ be the transport to A^* of the natural pairing on $A \times A'$. ∎

Definition 6.3.5. *Let M be a structure of finite rank, A a group interpretable in M without parameters.*

1. *Let S, T be definable sets. We write $S \subseteq^* T$ if $rk(S - T) < rk\, S$. For corresponding definable formulas σ, τ we use the notation $\sigma \Longrightarrow^* \tau$.*
2. *If B is a subgroup of A^*, and $a \in A$, then $gtp(a/B)$ denotes the atomic type of a over B in the language containing only the bilinear map $A \times A^* \to \mathbb{Z}/n\mathbb{Z}$, with n the exact exponent of A.*
3. *The group A is settled if for every algebraically closed parameter set C and $a \in A$ of maximal rank over C, we have $tp(a) \cup gtp(a/A^* \cap C) \Longrightarrow^* tp(a/C)$.*
4. *The group A is 2-ary if for any algebraically closed parameter set C and any set $\mathbf{b} = b_1, \ldots, b_n$ in A of elements which are independent over C of maximal rank, we have*

$$\bigcup_i tp(b_i/C) \cup \bigcup_{ij} tp(b_i b_j/\, acl\, \emptyset) \Longrightarrow^* tp(\mathbf{b}/C).$$

Our primary objective in the long run is to show that every group becomes both settled and 2-ary after introducing finitely many constants. The linear part of a quadratic geometry is an example of an unsettled group.

We close this section with a few miscellaneous lemmas.

Lemma 6.3.6. *Let $\mathcal{M}$ be a Lie coordinatizable structure and A, B groups 0-definably interpreted in $\mathcal{M}$ with no proper 0-definable subgroups of finite index. Suppose that B is settled. If a, b, c are independent, with $a \in A$ and $b \in B$ of maximal rank, then*

$$tp(b/a, acl(\emptyset)) \cup tp(b/\, acl(c)) \Longrightarrow^* tp(b/a, c).$$

Proof. As B is settled, taking $C = acl(a, c)$ we get

$$tp(b/\, acl(\emptyset)) \cup gtp(b/\, acl(a, c) \cap B^*) \Longrightarrow^* tp(b/a, c).$$

We will check that

$$(*) \qquad tp(b/a, acl(\emptyset)) \cup tp(b/\, acl(c)) \Longrightarrow gtp(b/\, acl(a, c) \cap B^*)$$

Let $d \in acl(a, c) \cap B^*$. We will apply Lemma 6.2.8 with: $A_1 = B^*$; $A_2 = A$; $C = \{c\} \cup [acl(c) \cap B^*]$; $a_2 = a$. To do so, we must work over $acl(\emptyset)$, noting that there are no $acl(\emptyset)$-definable proper subgroups of A or B of finite index, and thus, in particular, $acl(\emptyset) \cap B^* = (0)$. Thus by Lemma 6.2.8,

$$d = d_a + d_c \text{ with } d_a \in dcl(a, acl(\emptyset)) \cap B^* \text{ and } d_c \in acl(c) \cap B^*.$$

Thus $(*)$ holds. ∎

Lemma 6.3.7. *Let $\mathcal{M}$ be a Lie coordinatizable structure and suppose that A_i $(1 \le i \le n)$ is a family of groups 0-definable in $\mathcal{M}^{\mathrm{eq}}$, each having no 0-definable subgroups of finite index, and all but the first settled.*

Let C be algebraically closed and let $a_i, b_i \in A_i$ have maximal rank with $a_1, b_1, \ldots, a_n, b_n$ independent over C. If $tp(a_i/C) = tp(b_i/C)$ for all i and $tp(a_i, a_j / acl(\emptyset)) = tp(b_i, b_j / acl(\emptyset))$ for all i, j, then $tp(\mathbf{a}/C) = tp(\mathbf{b}/C)$.

Proof. We proceed inductively. Thus we may suppose

$$tp(a_1, \ldots, a_{n-1}/C) = tp(b_1, \ldots, b_{n-1}/C)$$

and even that $a_i = b_i$ for $i < n$. Let $A = A_1 \times \cdots \times A_{n-1}$ and apply the previous lemma to the pair A, B_n.   ∎

Corollary 6.3.8. *Let $\mathcal{M}$ be a Lie coordinatizable structure and A a 0-definable settled group in $\mathcal{M}^{eq}$ such that $acl(\emptyset) \cap A^* = (0)$. Then A is 2-ary.*

Note that the property that A is 2-ary will persist over a larger set of parameters, though the hypothesis will not necessarily persist.

Lemma 6.3.9. *Let $\mathcal{M}$ be a Lie coordinatizable structure, A and B 0-definable groups of exponent n with no 0-definable subgroups of finite index. Let $D \subseteq (A \times B)$ be a type over $acl(\emptyset)$ of maximal rank. Then the following are equivalent:*

1. *For $(a, b) \in D$, a lies in every b-definable subgroup of A of finite index.*
1'. *For $(a, b) \in D$, b lies in every a-definable subgroup of A of finite index.*
2. *For $(a, b) \in D$, there are a_1, a_2, a_3 in A with $a_1 + a_2 = a_3$, all realizing $tp(a/acl(b))$, and with a_1, a_2, b independent.*
3. *There are a_1, a_2, a_3 in A and $b \in B$ such that $(a_1, b) \in D$, with $a_1 + a_2 = a_3$, and*
$$tp(a_2 b / acl(\emptyset)) = tp(a_3 b / acl(\emptyset)).$$
4. *Every $acl(\emptyset)$-definable bilinear map $A \times B \to \mathbb{Z}/n\mathbb{Z}$ vanishes on D.*

Proof. (1) implies (2): Let $(a, b) \in D$, and let A^b be the smallest b-definable subgroup of A of finite index. Then $a \in A^b$. Let D' be the locus of a over $acl(b)$. Working over $acl(b)$, Lemma 6.1.4 applies. Thus the stabilizer $Stab(D')$ is a b-definable subgroup of A^b of finite index, and $Stab(D') = A^b$. Let $a_3 = a$. As $rk[Stab(D') - Stab_\circ(D')] < rk A^b$, we can find $a_2 \in D' \cap Stab_\circ(D')$ independent from a_3, b, and let $a_1 = a_3 - a_2$.

Evidently (3) is a weakening of (2). We show next that (3) implies (4). Let $f : A \times B \to \mathbb{Z}/n\mathbb{Z}$ be $\mathbb{Z}$-bilinear and algebraic over $acl(\emptyset)$. As D represents a complete type over $acl(\emptyset)$, f is constant on D; let the value be u. Then $f(a_2, b) = f(a_3, b) = f(a_1, b) + f(a_2, b)$ so $u = f(a_1, b) = 0$.

Since condition (4) is symmetric in A and B it suffices now to show that (4) implies (1). Assume condition (1) fails: $(a, b) \in D$, H is a b-definable subgroup of A of finite index, and $a \notin H$. Fix $f \in A^*$ vanishing on H with $f(a) \neq 0$. Note that $f \in acl(b)$. Let D^* be the locus of (f, b) over $acl(\emptyset)$, and S the stabilizer of D^* in $A^* \times A$. As f is algebraic over b, $S \cap [A^* \times (0)]$ is finite, and thus lies in $acl(\emptyset) \cap A^* = (0)$ by the condition on A. Furthermore $rk\, S = rk\, B$, and thus S projects onto B and is the graph of a homomorphism $h : B \to A^*$. D^* lies in the coset $S + (f, b) = S + (f - h(b), 0)$. Now the representative $f - h(b) \in acl(\emptyset) \cap A^* = (0)$ so $f = h(b)$. Define $(x, y) = [h(y)](x)$; then $(a, b) = f(a) \neq 0$. Thus (4) fails.   ∎

6.4 RANK AND MEASURE

We can attempt to construct a measure on subsets of a group A by taking cosets of a subgroup of index n to have measure $1/n$. Thus we may assign to a set S the infimum of the sums $\sum_i 1/n_i$ corresponding to coverings of S by finitely many such cosets. Our objective here is to show that the "measure zero" sets are those of less than full rank.

Lemma 6.4.1. *Let $\mathcal{M}$ be a Lie coordinatizable structure and A an abelian group of exponent p, 0-definably interpretable in $\mathcal{M}$. Let D be a 0-definable subset of A of full rank, and $a_1^*, \ldots, a_n^* \in A^*$ independent generics. Let $\alpha_1, \ldots, \alpha_n$ be elements of the prime field F_p. Then $\{d \in D : (d, a_i^*) = \alpha_i\}$ has full rank.*

Proof. By induction and the addition of parameters this reduces to the case $n = 1$. If this fails, then for $a_* = a_n$ a generic element of A^*, the complement D' of D contains a coset C_{a^*} of $\ker(a^*)$, modulo a set of smaller rank. We will argue that $rk\, D < rk\, A$.

Fix m, and let $b_1, \ldots, b_m$ be independent conjugates of a^*. We will consider the cardinality of D and of other definable sets in large envelopes E of $\mathcal{M}$. We have $|C_{a^*}| = q^{-1}|A|$ for some fixed q. The b_i are linearly independent in A^*, so $b_1, \ldots, b_m$ maps A onto F_p^m and the C_{b_i} are statistically independent. Thus the complement of $\bigcup_i C_{b_i}$ has cardinality $(1 - q^{-1})^m |A|$. Now $C_{b_i} \cap D$ has rank less than $rk\, A$, so in the limit $|C_{b_i} \cap D|/|A| \to 0$ by Lemma 5.2.6. Thus $\limsup_E |D|/|A| \leq (1 - q^{-1})^m$; varying m, $\lim_E |D|/|A| = 0$ and $rk\, D < rk\, A$ by Lemma 5.2.6.   ∎

Lemma 6.4.2. *Let $\mathcal{M}$ be Lie coordinatizable, let A be an abelian group interpreted in $\mathcal{M}$, and let $D \subseteq A$ be definable with $rk\, D = rk\, A$. Then finitely many translates of D cover A. More specifically, if D is c-definable then one may find $\mathbf{b} = (b_1, \ldots, b_n)$ in A with $A = \bigcup_i (D + b_i)$ and $\mathbf{b}$ independent from c.*

Proof. We may suppose that A is 0-definable, and we proceed by induction on the maximal length of a chain of $acl(\emptyset)$-definable subgroups of A.

We claim first that the result holds when A is part of a basic linear geometry for $\mathcal{M}$. We leave this essentially to the reader, but as an example, suppose A is an orthogonal space with quadratic form Q and $D = \{x \neq 0 : Q(x) = 0\}$. Let $V \leq A$ be nondegenerate of dimension 5. Then we claim $D + V = A$. Take $v \in A$, and choose w so that $\langle v, w \rangle$ is nondegenerate. Then $V_\circ = \langle v, w \rangle^\perp \cap V$ is a nondegenerate subspace of dimension at least 3, not containing v, and $Q(v) = Q(u)$ for some nonzero $u \in V_\circ$. Then $v - u \in D$.

Now suppose that A has a nontrivial $acl(\emptyset)$-definable finite subgroup B. Then $\bar{D} = (D + B)/B$ has full rank in A/B and induction applies to $\bar{D}$, A/B. As B is finite this yields the claim in A.

Assume now that A has no nontrivial $acl(\emptyset)$-definable finite subgroup, and is not part of a basic linear geometry. There is an $acl(\emptyset)$-definable subgroup A_1 of A which is part of a stably embedded basic linear geometry of $\mathcal{M}$ (Lemma 6.2.11). Let D be c-definable of full rank in A. Pick $b \in A$ of maximal rank over c such that $[b + A_1] \cap D$ is infinite. Then $D - b$ meets A_1 in an infinite set and thus there is a finite subset $F \subseteq A_1$ such that $A_1 \subseteq F + D - b$, and we may take the elements of F to be independent from b, c. Let B be the locus of b over $F \cup \{c\}$. Then B has full rank and for $b' \in B$, $A_1 \subseteq F + D - b'$. Now by induction in A/A_1, for some finite set F', $F' + B + A_1$ covers A. We claim that $F + F' + D = A$.

Let $a \in A$. Then for some $b' \in B$, we have $a \in F' + b' + A_1 \subseteq F' + b' + (F + D - b') = F' + F + D$, as claimed. ∎

Lemma 6.4.3. *Let $\mathcal{M}$ be Lie coordinatizable, let A be an abelian group interpreted 0-definably in $\mathcal{M}$, and suppose A has no proper 0-definable subgroups of finite index. Let $h_i : A \rightarrow B_i$ for $i = 1, 2$ be definable homomorphisms onto finite 0-definable groups B_1, B_2 and let $h = (h_1, h_2) : A \rightarrow B_1 \times B_2$ be the induced map. If h_1, h_2 are independent then h is surjective.*

Proof. Let the range of h be $C \leq B_1 \times B_2$ and let $C_1 = C \cap [B_1 \times (0)]$, $C_2 = C \cap [(0) \times B_2]$. C can be interpreted as the graph of an isomorphism between B_1/C_1 and B_2/C_2. Let $g_i : A \rightarrow B_i/C_i$ be the map induced by h_i. Then $g_i \in acl(h_i)$ and g_1 and g_2 differ only by an automorphism of the range. Thus $g_i \in acl(h_1) \cap acl(h_2) = acl(\emptyset)$ and thus by assumption $B_1 = C_1$, $B_2 = C_2$, and h is surjective. ∎

Lemma 6.4.4. *Let $\mathcal{M}$ be Lie coordinatizable, let A be an abelian group interpreted 0-definably in $\mathcal{M}$, let A^0 be the smallest 0-definable subgroup of finite index, and let $D \subseteq A$ be 0-definable with $rk\, D = rk\, A$. Assume that D lies in a single coset C of A^0 and let $h : A \rightarrow B$ be a definable homomorphism into a finite group B. Then for any $b \in h[C]$, D meets $h^{-1}[b]$ in a set of full rank.*

Proof. If h is algebraic over $\emptyset$ then h is constant on C and there is nothing to prove. Suppose, therefore, that $h \notin acl(\emptyset)$.

Using the previous lemma, the proof of Lemma 6.4.1 can be repeated (for the case $n = 1$), using independent conjugates of h. Alternatively, the following argument can be given which does not make use of finite approximations but again makes use of an infinite family of independent conjugates of h.

Let $\nu(h) = \nu_D(h) = |\{c \in h[C] : D \cap h^{-1}[c] \text{ has full rank }\}|/|h[C]|$. We claim $\nu(h) = 1$. For $h' = (h_1, h_2)$ induced by two homomorphisms, if $h'^{-1}[(c_1, c_2)] \cap D$ has full rank, then the same applies to $h_i^{-1}[c_i]$ and thus by the previous lemma, if h_1 and h_2 are independent then we get $\nu(h') \leq \nu(h_1)\nu(h_2)$. Thus if $\nu(h) < 1$, then by taking enough independent conjugates h_i of h we can construct a homomorphism f with finite image for which $\nu(f)$ is arbitrarily small. But a finite number m of translates $D + a_i$ cover C, and $\nu_D = \nu_{D+a}$ for each translate. Hence $1 = \nu_C(f) \leq m\nu_D(f)$, and we have a lower bound on ν_D, a contradiction. ∎

Lemma 6.4.5. *Let $\mathcal{M}$ be Lie coordinatizable, let A be an abelian group interpreted 0-definably in $\mathcal{M}$, and let D be the locus of a complete type over $\mathrm{acl}(\emptyset)$ of maximal rank. Then there are independent $a, a' \in D$ such that $a - a'$ lies in every a-definable subgroup of A of finite index.*

Proof. Take $a \in D$. Let A^a be the smallest a-definable subgroup of A of finite index. We consider the canonical homomorphism $h : A \to A/A^a$. The previous lemma applies and shows that $(A^a + a) \cap D$ has full rank. It suffices to take a' in the intersection, of maximal rank. ∎

6.5 THE SEMI-DUAL COVER

It is remarkable that duality can be used to reduce many aspects of the treatment of affine covers to the treatment of finite covers. (Affine covers are covers with affine fibers in the sense of §4.5, corresponding, for us, to stages in a Lie coordinatization in which affine geometries are involved.)

Suppose that $\pi : \mathcal{N} \to A$ is a cover with affine fibers $N_a = \pi^{-1}[a]$, affine over A. (Some might prefer to call A "V" here, but as in the previous section we tend to call our abelian groups A for the present.) Then the affine dual N_a^* is a finite cover of the linear dual A^*. Let $\mathcal{N}^*$ be the corresponding cover; then it seems that $\mathcal{N}^*$ should contain the same information as $\mathcal{N}$. We show below that a group structure on $\mathcal{N}$ corresponds to what we call a "bilinear group structure" on $\mathcal{N}^*$. This approach will lead to our sharpest result on groups, the "finite basis theorem" for definability in definable groups. Cf. the work of Ahlbrandt and Ziegler in [AZ2].

On the other hand, this method does not appear to apply to iterated covers, as a cover of $\mathcal{N}$ does not appear to correspond to a cover of $\mathcal{N}^*$, and thus the use of affine covers cannot be eliminated systematically.

THE SEMI-DUAL COVER**127**

Definition 6.5.1. *Let A_1, A_2 be abelian groups. A bilinear cover of A_1, A_2 is a surjective map $\pi = (\pi_1, \pi_2) : L \to A_1 \times A_2$, where L is a structure with two partial binary operations $q_1, q_2 : L \times L \to L$, with the following properties:*

BL1. *q_i is defined on $\cup_{a \in A_i}[\pi_i^{-1}[a] \times \pi_i^{-1}[a]]$, and gives an abelian group operation on each subset $L[a] = \pi_i^{-1}[a]$.*

BL2. *For $i, i' = 1, 2$ in either order, $\pi_{i'}$ is a group homomorphism on each group $(L[a]; q_i)$ for $a \in A_i$.*

BL3. *Given elements $a_{ij} \in A_i$ for $i = 1, 2$, $j = 1, 2$, and elements $c_{ij} \in \pi^{-1}(a_{1i}, a_{2j})$, we have*

$$q_2(q_1(c_{11}, c_{12}), q_1(c_{21}, c_{22})) = q_1(q_2(c_{11}, c_{21}), q_2(c_{12}, c_{22})).$$

In (BL3), note that the result of the calculation, carried out on either side, lies in $\pi^{-1}(a_{11} + a_{12}, a_{21} + a_{22})$.

Such covers will normally occur interpreted within some $\mathcal{M}^{eq}$, in which case L and all the associated structure is taken to be interpretable in $\mathcal{M}$. Generally, q_1 and q_2 will be given the more suggestive notations "$+^1$, $+^2$," or just "$+$" if no ambiguity results. The same applies to iterated sums $\sum^1$, $\sum^2$, or $\sum$. We will also write $L(a_1, a_2)$ for $\pi^{-1}[(a_1, a_2)]$.

Lemma 6.5.2. *Let $\pi : L \to A_1 \times A_2$ be a bilinear cover relative to the operations q_1 and q_2. Then:*

1. *q_1 and q_2 agree on $L(0, 0)$.*
   Let this group be denoted $(A, +)$.
2. *If $0_1, 0_2$ are the identity elements of A_1 and A_2 respectively, then there are canonical identifications $L(0_1) \simeq A \times A_2$ and $L(0_2) \simeq A_1 \times A$.*
3. *Each set $L(a_1, a_2)$ is naturally an affine space over $L(0, a_2)$, and also over $L(a_1, 0)$, giving two A-affine structures on $L(a_1, a_2)$ which coincide.*

Proof.
Ad 1. Let $A = L(0, 0)$ as a set. Let e_1, e_2 be the 0-element of A with respect to q_1 and q_2 respectively. With all a_{ij} equal to 0 (in A_1 or A_2, as the case may be) and with c_{ij} equal to e_1 in (BL3), and setting $e' = q_2(e_1, e_1)$, condition (BL3) can be written as $e' = q_1(e', e')$. Hence we have $q_2(e_1, e_1) = e' = e_1$, and this implies $e_1 = e_2$.

Then with $c_{12} = c_{21} = e_1$ we get $q_2 = q_1$ on A. We note in passing that with $c_{11} = c_{22} = e_1$ we would also get the commutative law (or laws) on A, which in any case we have assumed.

Ad 2. We now consider the structure of $L(0_1)$. By (BL2) we have

$$q_2[L(a_1, a_2), L(a_1', a_2)] \subseteq L(a_1 + a_1', a_2)$$

and, in particular, $L(0_1, a_2)$ is a subgroup of $L(a_2)$ for $a_2 \in A_2$. Let its identity element be denoted $z(a_2)$. We will show that $z : A_2 \to L(0_1)$ is a homomorphism. Let $a, a' \in A_2$ and let $z = q_1(z(a), z(a'))$. Applying (BL3), we get $q_2(z, z) = q_1(q_2(z(a), z(a)), q_2(z(a'), z(a'))) = q_1(z(a), z(a')) = z$ and thus $z = z(a + a')$. Thus z is a homomorphism. By definition $\pi_2 z$ is the identity and as the kernel of π_2 on $L(0_1)$ is the group A, we get a direct product decomposition $(L(0_1), q_1) \simeq A_2 \times A$. This identification respects π_2; that is, $L(0_1, a)$ corresponds to $(a_2) \times A$ with q_2 acting on $L(0_1, a)$ as on A.

A similar analysis applies on the other side.

Ad 3. According to (BL2) under q_2 $L(0, a_2)$ acts on $L(a_1, a_2)$ for any $a_1 \in A_1$, making the latter an affine space over the former. After identifying A with $L(0_1, a_2)$ and $L(a_1, 0_2)$ we get two affine actions of A on $L(a_1, a_2)$. These can be compared as follows. Let $x \in A$, $y \in L(a_1, a_2)$, and let z_1 be the identity element of $L(a_1)$, z_2 the identity element of $L(a_2)$. The identification of A with $L(0_1, a_2)$ takes x to $q_1(x, z_2)$; the other identification takes x to $q_2(x, z_1)$. For the action of A via $L(0_1, a_2)$ on $L(a_1, a_2)$ we get $q_2(q_1(x, z_2), y) = q_2(q_1(x, z_2), q_1(z_1, y)) = q_1(q_2(x, z_1), q_2(z_2, y)) = q_1(q_2(x, z_1), y)$, which is the action of A via $L(a_1, 0_2)$. ∎

Lemma 6.5.3. *Let L be a bilinear cover of $A_1 \times A_2$. Let $a_i \in A_1$, $a'_i \in A_2$, and let $x_{ij} \in L(a_i, a'_j)$, r_i, s_j integer coefficients. Then $\sum_i^2 r_i \sum_j^1 s_j x_{ij} = \sum_j^1 s_j \sum_i^2 r_i x_{ij}$ and, in particular, if $r_i = s_j = 1$ then the order of summation can be reversed.*

Proof. We first deal with the case in which $r_i = s_j = 1$, proceeding by induction on the numbers m, n of indices i and j respectively, beginning with $m = n = 2$, which is (BL3). Case $(m, n + 1)$ is easily derived from cases (m, n) and $(m, 2)$ as in the usual proofs of basic properties of sums, and case $(m + 1, n)$ follows similarly from (m, n) and $(2, 2)$, so from the basic case $m = 2, n = 2$ we can first get case $(m, 2)$ for any m and then (m, n) for any m, n.

The general case of integer coefficients follows by simply expanding out the definitions from the case of coefficients ± 1. So consider now the case in which the r_i are ± 1, but keep the $s_j = 1$. Splitting the set I of indices i into I^+ and I^- according to the sign of r_i, our claim is

$$\sum_{I^+}^2 \left(\sum_j^1 x_{ij} \right) -^2 \sum_{I^-}^2 \left(\sum_j^1 x_{ij} \right) = \sum_j^1 \sum_i^2 r_i x_{ij}.$$

Moving the negative term from left to right and applying the positive case twice, with a little care, the claim falls out. The case of $r_i, s_j = \pm 1$ then follows by repeating the argument. ∎

Lemma 6.5.4. *Let $\mathcal{M}$ be a structure, and*

$$0 \to A_1 \to B \to A_2 \to 0$$

be an exact sequence of abelian groups with A_1, A_2 of prime exponent p, and assume this sequence is interpreted in $\mathcal{M}$. For $a \in A_2$ let B_a be the preimage in B of a, a coset of A_1, and let B_a^* be the set of definable affine homomorphisms from B_a to the field F of p elements. Let $L = \{(a, f) : a \in A_2, f \in B_a^*\}$, take $\pi_1 : L \to A_2$ natural, and let $\pi_2 : L \to A_1^*$ be defined by $\pi_2 f \in A_1^*$ the linear map associated to f, i.e. $f(x + y) - f(y)$ as a function of x.

Then L is a cover of $A_2 \times A_1^*$ with respect to the operations q_1 and q_2 described as follows. The operation q_1 acts by addition in the second coordinate. The operation q_2 also acts by addition, but in a somewhat more delicate sense: if $\pi_2(a, f) = \pi_2(a', f')$ then f and f' are affine translates of the same linear map $f_\circ$, and we set $q_2((a, f), (a', f')) = (a + a', f + f')$ where $f + f'$ is the function g on $B_{a+a'}$ defined by $g(b + b') = f(b) + f'(b')$ for $b \in B_a$, $b' \in B_{a'}$.

Proof. One checks in the first place that q_2 is well defined: for $a_1 \in A_1$, $f(b + a_1) + f'(b' - a_1) = f(b) + f_\circ(a_1) + f'(b') - f_\circ(a_1) = f(b) + f'(b')$.

The verification of the axioms is straightforward. Axiom (BL3) concerns the situation $a, a' \in A_2$, $f_1, f_2 \in B_a^*$, $f_1', f_2' \in B_{a'}^*$, f_1 and f_2 induce the same linear map, and f_1' and f_2' induce the same linear map. The result of applying the appropriate combinations of q_1 and q_2 in either order is $(a + a', (f_1 + f_2) + (f_1' + f_2'))$ with the sum on the right involving $B_{a+a'} = B_a + B_{a'}$. ∎

The cover associated to an exact sequence as described above will be called a *semi-dual* cover since it involves two groups, one of which is a dual group. Notice that the "structure group" $L(0, 0)$ for the semi-dual cover associated with such an exact sequence is the set of constant maps from A_1 to F, which we identify with F. If $\mathcal{M}$ is Lie coordinatized then the cover obtained is definable since the dual group is definable.

Now we present a construction in the reverse direction.

Lemma 6.5.5. *Let $\mathcal{M}$ be a structure, A_1 and A_2 abelian groups interpreted in $\mathcal{M}$, and L a bilinear cover of $A_2 \times A_1$ interpreted in $\mathcal{M}$. Set $F = L(0, 0)$, and let B be the set*

$$\{(a, f) : \quad a \in A_2, f : L(a) \to F \text{ definable},$$
$$f \text{ is the identity on } L(a, 0) \text{ identified with } L(0, 0)\}.$$

Then B is a group with respect to the operation $(a, f) + (a', f') = (a + a', f'')$ with $f''(q_2(x, x')) = f(x) + f(x')$ for $x \in L(a)$, $x' \in L(a')$ and $\pi_2(x) = \pi_2(x')$, and there is an exact sequence

$$0 \to \mathrm{Hom}(A_1, F) \to B \to A_2 \to 0$$

where Hom is the group of definable homomorphisms.

Proof. Wherever one sees an expression $q_2(x, x')$ it should be assumed that $\pi_2(x) = \pi_2(x')$, both in the above and in the proof following.

We check first that the operation $+$ on B is well defined. Let $x, y \in L(a)$, $x', y' \in L(a')$, with $q_2(x, x') = q_2(y, y')$. We may write $y = q_1(x, \alpha)$, $y' = q_1(y, \beta)$, with $\alpha \in L(a, 0)$, $\beta \in L(a', 0)$ (or $\alpha, \beta \in L(0, 0)$ after appropriate identifications). The relation $q_2(x, x') = q_2(y, y')$ after application of (BL3) becomes $q_1(\alpha, \beta) = 0$ or $\alpha + \beta = 0$ in $L(0, 0)$. Thus $f(y) + f'(y') = f(x) + \alpha + f'(x') + \beta = f(x) + f'(x')$ as needed.

The operation $+$ is clearly commutative and associative, and one can easily construct inverses. Thus we have a group B, and a projection from B to A_2. The kernel is $\{(0, f) : 0 \in A_2, f : L(0) \to F, f$ is the identity on $F\}$. But $L(0)$ can be identified with $A_1 \times F$ and thus this kernel can be identified with the definable homomorphism group $Hom(A_1, F)$. ∎

Definition 6.5.6. *An abelian group A of prime exponent interpreted in a Lie coordinatized structure will be called* reflexive *if the natural map $A \to A^{**}$ is an isomorphism.*

Lemma 6.5.7. *Let $\mathcal{M}$ be a Lie coordinatizable structure, A an abelian group interpreted in $\mathcal{M}$. Then the following are equivalent:*

1. *A is reflexive.*
2. *The natural map $A \to A^{**}$ is injective.*
3. *A is definably isomorphic to a dual group B^*.*

Proof. (2) implies (1): As in the proof of Proposition 6.3.2, using finite approximations to compare cardinalities, we get $|A^{**}| \leq |A^*| \leq |A|$.

Evidently (3) implies (2) and (1) implies (3). ∎

Lemma 6.5.8. *Let $\mathcal{M}$ be a Lie coordinatized structure, and A_1, A_2 abelian groups interpreted in $\mathcal{M}$ of prime exponent p, with A_1 reflexive. Let F be the field of order p. Then there is a natural correspondence between interpretable exact sequences $0 \to A_1 \to B \to A_2 \to 0$ and definable bilinear covers L of $A_2 \times A_1^*$ with structure group $L(0, 0) = F$, up to the natural notions of isomorphism.*

Proof. This is largely contained in Lemmas 6.5.4 and 6.5.5, bearing in mind that the groups A_1, A_2 of Lemma 6.5.5 are A_1^* and A_2 in our present notation. It is also necessary to trace through the claim that these two correspondences reverse one another up to canonical isomorphism, a point which we leave to the reader. ∎

The next proposition (after a preparatory lemma) states essentially that definable sections of bilinear covers are locally affinely bilinear, uniformly in a parameter: on a complete type, they respect the bilinear structure, up to translation. It would be interesting to get a global analysis. The proof requires that one of the groups be settled, a hypothesis which will eventually be seen to hold generally over an appropriate set of parameters; but the proof of the latter result requires the present one.

Notation 6.5.9

1. For $D \subseteq A \times B$, $s : A \times B \to C$, and $a \in A$, we write D_a for $\{b \in B : (a, b) \in D\}$ and $s_a : D_a \to C$ for the map induced by s.

2. For A an $\aleph_0$-categorical group, c a parameter or finite set of parameters, let A^c be the smallest c-definable subgroup of A of finite index. This will be called the principal component of A over c.

Notice the law

$$(A_1 \times A_2)^c = A_1^c \times A_2^c$$

and hence $(A^n)^c = (A^c)^n$.

Lemma 6.5.10. Let $\mathcal{M}$ be Lie coordinatizable, A and B abelian groups and $\pi : L \to A \times B$ a bilinear cover, all 0-definably interpreted in $\mathcal{M}$, with structure group $F = L(0,0)$. Let $f : A' \to A$ be a generically surjective 0-definable map, $D \subseteq A' \times B$ the locus of a complete type over $acl(\emptyset)$ of maximal rank, and $s : D \to L$ a 0-definable section relative to f, i.e. $s(a', b) \in L(fa', b)$ on D. Assume

1. The group B is settled.
2. A and B have no 0-definable proper subgroups of finite index.
3. $acl(a') \cap B^* = dcl(a') \cap B^*$ for $a' \in A'$.
4. For $(a', b) \in D$, b lies in $B^{a'}$, the principal component of B over a'.

Then for any $a' \in A'$, the map $s_{a'} : D_{a'} \to L(fa')$ is affine; that is, it is induced by an affine map.

Proof. We may work over $acl(\emptyset)$. As B is settled it follows from (3) that $D_{a'}$ is the locus of a complete type over $acl(a')$.

Let D^* be

$$\{(a', b_1, b_2, b_3, b_4) : \quad \text{the first four coordinates are independent,}$$
$$\text{all } (a', b_i) \text{ lie in } D, \text{ and } b_4 = b_1 - b_2 + b_3\}.$$

By Lemma 6.2.7 it suffices to check the relation

$$s(a', b_4) = s(a', b_1) - s(a', b_2) + s(a', b_3)$$

on D^*.

Fix (a', b_1, b_2, b_3, b_4) in D^*. We claim that there are elements a'_1, a'_2, a'_3 in A' such that

(i) $fa'_3 = fa'_1 + fa'_2$;

(ii) $tp(a'_1 a'_2 a'_3 b_i)$ does not depend on $i = 1, \ldots, 4$;

(iii) $tp(a'_i, \mathbf{b}) = tp(a', \mathbf{b})$ for $i = 1, 2, 3$.

Granted this, we may complete the computation as follows. Set $\alpha_i = \sum_j^1 (-1)^j s(a'_i, b_j)$, and $\beta_j = \sum_i^2 (-1)^i s(a'_i, b_j)$. By Lemma 6.5.3 we have

$$\sum_i^2 (-1)^i \alpha_i = \sum_j^1 (-1)^j \beta_j. \text{ As } \sum_j (-1)^j b_j = 0, \ \alpha_i \in L(a_i, 0_B). \text{ Let}$$

$\theta : L(0_B) \rightarrow A \times F$ be the canonical isomorphism: $\theta(x) = (a, \theta_2(x))$ for $x \in L(a, 0_B)$. Since we are working over $acl(\emptyset)$, $\theta_2 : L(a, 0_b) \rightarrow F$ is constant on the α_i, by condition (iii). Set $\alpha = \theta_2(\alpha_i)$. Thus

$$\theta_2 \left(\sum_i^2 (-1)^i \alpha_i \right) = (0, -\alpha).$$

Similarly, $\beta_j \in L(0_A, b_j)$ and under the isomorphism $\psi : L(0_A) \rightarrow B \times F$ we get $\psi(\beta_j) = (b_j, \beta)$ for a fixed β, and thus

$$\psi \left(\sum_j^1 (-1)^\beta_j \right) = (0, 0).$$

But ψ and θ agree on $L(0, 0)$, so the last two computations yield $\alpha = 0$, $\theta(\alpha_i) = 0$ in $L(0_B)$ and hence also in $L(a_i)$; that is, $\sum_j^1 s(a'_i, b_j) = 0$, as required.

It remains to choose the elements a_1, a_2, a_3. Let $a = fa'$. Each b_i is in $B^{a'}$, and hence $(b_1, b_2, b_3) \in (B^3)^a$. By Lemma 6.3.9 there are a_1, a_2, a_3 in A with $a_1 + a_2 = a_3$ such that a_1, a_2, b_1, b_2, b_3 are independent and all a_i realize $tp(a/ \, acl(b_1, b_2, b_3))$. Again by Lemma 6.3.9, each a_i lies in A^b for $\mathbf{b} = (b_1, b_2, b_3)$ and thus $(a_1, a_2) \in (A^2)^b$, and again by Lemma 6.3.9 $\mathbf{b} \in (B^3)^{a_1, a_2}$. As $a_3 \in dcl(a_1, a_2)$ we conclude

$$b_i \in (B^3)^a$$

with $\mathbf{a} = (a_1, a_2, a_3)$, for $i = 1, 2, 3$, but also for $i = 4$, as $(B^3)^a$ is a group.

Choose elements $a'_i \in A'$ above a_i for $i = 1, 2, 3$ satisfying $tp(a'_i) = tp(a')$. These are not yet the desired elements. Choose $\mathbf{b}' = (b'_1, b'_2, b'_3) \in (B^3)^{a'_1, a'_2, a'_3}$ with $tp\,\mathbf{b}' = tp\,\mathbf{b}$ and $rk\,\mathbf{b}' = 3\,rk\,B$. This is possible by Lemma 6.4.4 applied to B^3.

As B is settled $tp(b'_i/a_1 a_2 a_3) = tp(b_i/a_1 a_2 a_3)$. By the corollary to Lemma 6.3.7 B is 2-ary and thus $tp(b'_1 b'_2 b'_3/a_1 a_2 a_3) = tp(b_1 b_2 b_3/a_1 a_2 a_3)$. Applying an automorphism, we may suppose $b'_i = b_i$ for $i = 1, 2, 3$; this gives new values of a_i'. Condition (i) is satisfied, and as B is settled and $b_i \in B^{a_1', a_2', a_3'}$ for all i, condition (ii) is also satisfied. Finally, as B is settled and 2-ary we get condition (iii) as well. ∎

The next proposition is the preceding lemma with its fourth hypothesis deleted.

Proposition 6.5.11. *Let $\mathcal{M}$ be Lie coordinatizable, A and B abelian groups and $\pi : L \to A \times B$ a bilinear cover, all 0-definably interpreted in $\mathcal{M}$, with structure group $F = L(0,0)$. Let $f : A' \to A$ be a generically surjective 0-definable map, $D \subseteq A' \times B$ the locus of a complete type over $acl(\emptyset)$ of maximal rank, and $s : D \to L$ a 0-definable section relative to f, i.e. $s(a', b) \in L(fa', b)$ on D. Assume*

1. *The group B is settled.*
2. *A and B have no 0-definable proper subgroups of finite index.*
3. *$acl(a') \cap B^* = dcl(a') \cap B^*$ for $a' \in A'$.*

Then for any $a' \in A'$, the map $s_{a'} : D_{a'} \to L(fa')$ is affine, that is, is induced by an affine map.

Proof. As in the previous argument we work over $acl(\emptyset)$.

In the notation of the preceding proof, our claim is this: for (a', b_1, b_2, b_3, b_4) in D^*, we have $\sum_j (-1)^j s(a', b_j) = 0$. We claim first that there is $b \in D_{a'}$, independent from a', b_1, b_2, b_3, b_4, with $b - b_i \in B^{a', b}$ for $i = 1, 2, 3, 4$. As the principal component is a group, it suffices to deal with b_i for $i \leq 3$. As $D_{a'}$ is the locus of a complete type over the algebraically closed set $acl(a')$, by the type amalgamation property it will suffice to deal with a single b_i. This case is covered by Lemma 6.4.5.

Now let a'' enumerate $[acl(a', b) \cap B^*] \cup \{a', b\}$ and let f_1, f_2 be definable functions picking out a', b, respectively, from a''. Let A'' be the locus of a'' and let f'' be $f \circ f_1$. Let D' be the locus of $(a'', b_i - b)$. As B is settled, this set does not depend on i. Define $s' : D' \to L$ by $s'(x, u) = s(f_1 x, u + f_2 x) - s(f_1 x, f_2 x)$ with the subtraction performed in $L(f_1 x)$. Then in the context of A'', D', s', hypothesis (3) again holds, and hypothesis (4) of the preceding lemma is achieved. Thus $s'_{a''}$ is affine. Furthermore, each $b_i - b$ lies in $D'_{a''}$, so we get

$$
0 = \sum_i (-1)^i s'(a'', b_i - b) = \sum_i (-1)^i [s(a', b_i) - s(a', b)]
$$
$$
= \sum_i (-1)^i s(a', b_i)
$$

as claimed. ∎

6.6 THE FINITE BASIS PROPERTY

Our objective in the present section is to pin down definability in groups rather thoroughly, as follows.

Proposition 6.6.1 (Finite Basis Property). *Let M be Lie coordinatizable and A an abelian group interpreted in M. Then there is a finite collection of definable subsets D_i of A such that every definable subset of A is a boolean combination of the sets D_i, cosets of definable subgroups of A of finite index, and sets of rank less than $rk(A)$.*

The proof will occupy most of this section.

Lemma 6.6.2. *Let M be Lie coordinatizable and A an abelian group interpreted in M. The following are equivalent:*

1. *A is settled over $\emptyset$; i.e., we have*

   $$(*) \qquad\qquad tp(a/\emptyset) \cup gtp(a/C \cap A^*) \implies^* tp(a/C)$$

   for $a \in A$ of maximal rank over the algebraically closed set C.
2. *For every finite set $C_\circ$ there is an algebraically closed set C containing $C_\circ$ such that for $a \in A$ of maximal rank over C the relation $(*)$ holds.*
3. *Every definable subset of A is a boolean combination of 0-definable sets, cosets of definable subgroups of finite index, and sets of rank less than $rk\,A$.*

Proof. (2) is a weakening of (1), of course, and it implies (3), taking $C_\circ$ to be a defining set of parameters for the given definable set. Thus we are concerned only with the implication from (3) to (1).

Suppose on the contrary the implication

$$(*) \qquad\qquad tp(a/\emptyset) \cup gtp(a/C \cap A^*) \implies^* tp(a/C)$$

fails to hold generically over some algebraically closed set C, which we may take to be finitely generated. Take a type p over C of full rank other than $tp(a/C)$, compatible with the data in $(*)$. Let D be the locus of p. Now D lies in a single coset X of the principal component A^C. By (3), the type $tp(a/C)$ contains the intersection of some definable coset with $tp(a/\emptyset)$ up to a set of smaller rank; that is, there is a definable homomorphism h from A to a finite group, and a value c of h, such that $tp(a/\emptyset) \cup \{h(x) = c\} \implies^* tp(a/C)$. Hence $rk(D \cap h^{-1}[c]) < rk\,A$, contradicting Lemma 6.4.4. ∎

Thus Proposition 6.6.1 is equivalent to the statement that every group becomes settled over some finite set.

Lemma 6.6.3. *Let M be a Lie coordinatizable structure, and let $A_1, \ldots, A_n$ be settled groups 0-definably interpreted in M, with no proper 0-definable subgroups of finite index. Then the product $A = \prod_i A_i$ is settled over $acl(\emptyset)$.*

Proof. We may assume $n = 2$ and $acl(\emptyset) = dcl(\emptyset)$. Let C be algebraically closed, and $a = (a_1, a_2) \in A = A_1 \times A_2$ of maximal rank over C. Note that $A^* = A_1^* \times A_2^*$ and $C \cap A^* = (C \cap A_1^*) \times (C \cap A_2^*)$. Our claim is

$$tp(a_1, a_2/\emptyset) \cup gtp(a_1/C \cap A_1^*) \cup gtp(a_2/C \cap A_2^*) \Longrightarrow^* tp(a/C).$$

We have $tp(a_2/\emptyset) \cup gtp(a_2/C \cap A_2^*) \Longrightarrow^* tp(a_2/C)$, so to conclude it will suffice to show that $tp(a_1/a_2) \cup gtp(a_1/C \cap A_1^*) \Longrightarrow^* tp(a_1/a_2C)$, which is not quite what we have assumed. As A_1 is settled we have, in fact,

$$tp(a_1/\emptyset) \cup gtp(a_1/ \, acl(a_2, C) \cap A_1^*) \Longrightarrow^* tp(a_1/a_2C)$$

so it remains to understand $gtp(a_1/ \, acl(a_2, C) \cap A_1^*)$.

We apply Lemma 6.2.8 to A_1^* and A_2. Thus $acl(a_2, C) \cap A_1^* = [dcl(a_2) \cap A_1^*] + [C \cap A_1^*]$. As $gtp(a_1/ \, dcl(a_2) \cap A_1^*)$ is determined by $tp(a_1/a_2)$, we are done.   ∎

Definition 6.6.4. *Let A be an abelian group interpreted in a Lie coordinatizable structure $\mathcal{M}$. A definable subset Q of A will be called* tame *if every definable subset of Q is the intersection with Q of a boolean combination of cosets of definable subgroups of finite index, and sets of lower rank. This notion is of interest only when $rk \, Q = rk \, A$.*

Lemma 6.6.5. *Let $\mathcal{M}$ be a Lie coordinatizable structure, and let A be an abelian group interpreted in $\mathcal{M}$.*

1. *If A contains a definable tame subset Q of full rank, then A is settled over some finite set.*
2. *If A contains a settled definable subgroup B of finite index then A is settled over some finite set.*

Proof.

Ad 1. By Lemma 6.4.2 A can be covered by finitely many translates of Q. It suffices to work over a set of parameters C containing defining parameters for Q together with sufficiently many parameters of translation to cover A.

Ad 2. This is a special case of the first part, taking Q to be the locus of a 1-type over $\emptyset$ of full rank in B.   ∎

Lemma 6.6.6. *Let $\mathcal{M}$ be a Lie coordinatizable structure, and let A be an abelian group interpreted in $\mathcal{M}$. If A contains a finite subgroup $A_\circ$ for which the quotient $A/A_\circ$ is settled over a finite set, then A is settled over a finite set.*

Proof. Let $A/A_\circ$ be settled over $C_\circ$. Take $a \in A$ of maximal rank over $C_\circ$ and let $\bar{a} = a + A_\circ$ viewed as an element of the quotient group. Then a is algebraic over $\bar{a}$. Take C containing $C_\circ$, independent from a, with the multiplicity of

$tp(a/\bar{a}, C)$ minimized. Let q be the type of a over C. We claim that the locus Q of q is tame, in other words that for C' containing C and independent from a over C, we have

$$q \cup gtp(a/\,acl(C') \cap A^*) \Longrightarrow^* tp(a/C').$$

In any case our choice of C ensures that

$$(*) \qquad\qquad tp(a/\bar{a}, C) \Longrightarrow tp(a/C').$$

Let q' be $tp(\bar{a}/C)$. As the quotient group is settled,

$$q' \cup gtp(\bar{a}/(A/A_\circ)^*) \Longrightarrow^* tp(\bar{a}/C).$$

Now $(A/A_\circ)^*$ may be identified with a definable subset of A^* and thus in view of $(*)$, $q \cup gtp(a/A^* \cap acl(C')) \Longrightarrow^* tp(a/C')$. Thus Q is tame and A is settled over some finite set. ∎

Lemma 6.6.7. *Let $\mathcal{M}$ be a Lie coordinatizable structure, and let A be an abelian group interpreted in $\mathcal{M}$, A_1 a rank 1 $acl(\emptyset)$-definable subgroup of A, and suppose $acl(\emptyset) \cap A^* = (0)$ (i.e. A has no 0-definable subgroup of finite index), and $acl(\emptyset) \cap A_1 = (0)$. Suppose a is an element of A of full rank over $\emptyset$, with $a \in acl(a/A_1, c)$ for some c independent from a/A_1 (an element of A/A_1). Then there is an $acl(\emptyset)$-definable subgroup A_2 with $A = A_1 \oplus A_2$.*

Proof. Let Q be the locus of a over $acl(c)$. With $n = rk\,A$, the hypotheses give $rk(a/c) = n - 1$. Let $S = Stab(Q)$. Then S is a subgroup of A of rank $n - 1$ (Lemma 6.2.5), and Q lies in a single coset of S. We claim that $S \cap A_1$ is finite.

If $S \cap A_1$ is infinite, let $b \in S \cap A_1$ have rank 1 over $\emptyset$. By Lemma 6.2.5, part (4), we may take $b \in Stab_\circ Q$. Then there is $a' \in Q$ of rank $n-1$ over b, c such that $a'' = a' - b \in Q$. Thus $tp(a''/c) = tp(a/c)$ and $a'' \in acl(a''/A_1, c)$; that is, $a' - b \in acl(a'/A_1, c)$ and hence $b \in acl(a', c)$. This contradicts the independence of a', b over c.

Now by Proposition 6.2.3 there is an $acl(\emptyset)$-definable subgroup A_2 commensurable with S. It follows easily that $A_1 \cap A_2 = 0$ and $A_1 \oplus A_2$ is a definable subgroup of A of finite index defined over $acl(\emptyset)$, and thus $A_1 \oplus A_2 = A$. ∎

Lemma 6.6.8. *Let $\mathcal{M}$ be a Lie coordinatizable structure, let*

$$(0) \to A_1 \to B \to A_2 \to (0)$$

be an exact sequence of abelian groups interpreted in $\mathcal{M}$, with A_2 settled over $\emptyset$, and let $\pi : L \to A_2 \times A_1^*$ be the corresponding bilinear cover. Assume that $acl(\emptyset) \cap A_1 = (0)$ and $acl(\emptyset) \cap A_2^* = (0)$. Let C be algebraically *closed, and let D be a complete type over C in A_2 of maximal rank. Let $a^* \in C \cap A_1^*$ be generic in A_1^* over $\emptyset$, and suppose $g : D \to L(a^*)$ is a C-definable section, that is: $g(a) \in L(a, g_2(a))$ for some function g_2; here we use the standard representation of the bilinear cover L, and, in particular, $g_2(a)$ induces a^* on A_1. Then there is a C-definable homomorphism j from A_2 to a finite group, such that for any $b \in B$ with $b/A_1 \in D$, the quantity*

$$[g_2(b/A_1)](b)$$

is determined by $j(b)$.

Proof. We apply Proposition 6.5.11 with the groups A_1^* and A_2 here playing the role of the A of B from that proposition. For A' we take the locus of C (as an enumerated set) over $\emptyset$ and for the D of Proposition 6.5.11 we take the locus of (C, d) with d a realization of the type D from the present Lemma. The function f picks out the element corresponding to our a^* in any realization of the type of the sequence C. In particular, in the notation of Proposition 6.5.11, our present C is a typical element a'. Now applying Proposition 6.5.11, the section g is affine. In other words, if A'_2 is the principal component A_2^C, then A'_2 is a C-definable subgroup of finite index in A_2, and there is a C-definable homomorphism $h : A'_2 \to L(a^*)$ such that for $d, d' \in D$ we have $g(d) - g(d') = h(d - d')$. We may write $h(a) = (h_1(a), h_2(a))$ and as g is a section we find $h_1(a) = a$.

Let $B' = \{b \in B : b/A_1 \in A'_2\}$. Define a map $j_\circ$ from B' to the prime field F by $j_\circ(b) = [h_2(b/A_1)](a)$. We will show that $j_\circ$ is a homomorphism.

As h is a homomorphism, $j_\circ(b + b')$ is the second component of $h(b/A_1) + h(b'/A_1)$, evaluated at $b + b'$; by the definition of the operation q_2 on L, this is $h_2(b/A_1)(b) + h_2(b'/A_1)(b') = j_\circ(b) + j_\circ(b')$.

Thus $j_\circ$ is a homomorphism. Let B'' be its kernel, and let j be the canonical homomorphism from B to B/B''. We claim that this j works. Suppose $b_1, b_2 \in B$, $b_i/A_1 \in D$, and $j(b_1) = j(b_2)$. Then $b_1 - b_2 \in B''$ and $j_\circ(b_1 - b_2) = 0$, so $g_2(b_1/A_1)(b_1) = g_2(b_2/A_1)(b_2)$ is determined by the value of j. ∎

Lemma 6.6.9. *Let $\mathcal{M}$ be a Lie coordinatizable structure, let A be 0-definably interpretable in $\mathcal{M}$, A_1 a definable subgroup, and suppose that A_1 is settled. Suppose there is a 0-definable type of full rank in A with locus Q such that for any C and any $a \in Q$ with a/A_1 of maximal rank over C,*

$$(*) \qquad tp(a/(a/A_1)) \cup gtp(a/\, acl(C) \cap A^*) \implies tp(a/(a/A_1), C).$$

Then Q is tame in A, and hence A is settled over some finite set.

Proof. Let $\bar{a} = a/A_1$, and let $q = tp(\bar{a}/C)$. Then

$$q \cup gtp(\bar{a}/\, acl(C) \cap (A/A_1)^*) \implies^* tp(\bar{a}/C).$$

As $(A/A_1)^*$ can be identified with a definable subset of A^*, this together with $(*)$ yields

$$tp(a/C) \cup gtp(a/\, acl(C) \cap A^*) \Longrightarrow^* tp(a/C).$$

Thus Q is tame. ∎

The following lemma is critical.

Lemma 6.6.10. *Let $\mathcal{M}$ be a Lie coordinatizable structure, let A be 0-definably interpretable in $\mathcal{M}$, with $acl(\emptyset) \cap A^* = (0)$, and let A_1 be a 0-definable subgroup of A which is part of a stably embedded linear geometry J in $\mathcal{M}$, not of quadratic type. Assume that A/A_1 is settled and that there is no $acl(\emptyset)$-definable complement to A_1 in A. Then A is settled over some finite set.*

Proof. We will arrive at the situation of the previous lemma, relative to some finite set of auxiliary parameters $C_\circ$ (so the sets C of the previous lemma should contain $C_\circ$). We work over $acl(\emptyset)$.

Let $\bar{A} = A/A_1$. Fix an element $a \in A$ of maximal rank, and let $\bar{a} = a/A_1$. Let $S = a + A_1$ viewed as an affine space over A_1. Let $S^{*\circ}$ be the prime field affine dual defined in §2.3. Call a set C *basal* if C is algebraically closed and independent from a. Then we claim

> For C basal, a is not in $acl(\bar{a}, C, J)$.

Otherwise, take $a \in acl(\bar{a}, C, d_1, \ldots, d_k)$ with $d_i \in J$ and k minimal. Then the sequence

$$\bar{a}, C, d_1, \ldots, d_k$$

is independent. We apply Lemma 6.6.7, noting that $acl(\emptyset) \cap A_1 = (0)$ by our hypothesis. Then Lemma 6.6.7 produces a complement to A_1 in A, a contradiction. Also, by Lemma 6.2.8 $acl(\bar{a}, C) \cap J = dcl(\bar{a}, C) \cap J$. Now Lemma 2.3.17 applies, giving

$$tp(a/\bar{a}, dcl(\bar{a}, C) \cap S^{*\circ}) \Longrightarrow tp(a/\bar{a}, C).$$

Let $T(C)$ be $dcl(C) \cap S^{*\circ}$. We need to examine $T(C)$ more closely for basal C. For $f \in A_1^*$ let $S^{*\circ}(f)$ be the set of elements of $S^{*\circ}$ lying above f; this is an affine space over the prime field $F_\circ$, of dimension 1. Let $A_1^*(C) = acl(C) \cap A_1^*$. Let $T_1(C) = dcl(C, \bar{a}) \cap \bigcup \{S^{*\circ}(f) : f \in A_1^*(C)\}$. We claim that for some basal C, for all C' containing C, we have

$(*)$ $$T(C') = T(C) + T_1(C')$$

and hence $T(C') \subseteq dcl(\bar{a}, T(C), T_1(C'))$.

Let $\beta(C) = \{x \in A_1^*(\bar{a}) : \text{for some } y \in A_1^*(C), S^{*\circ}(x + y) \cap T(C) \neq \emptyset\}$. Choose C basal with $\beta(C)$ maximal. Let $C' \supseteq C$ be basal, $t \in T(C')$. Then

$t \in S^{*\circ}(x + y)$ for some $x \in A_1^*(\bar{a})$, $y \in A_1^*(C')$. So $t \in \beta(C') - \beta(C)$. Thus there is $y' \in A_1^*(C)$ and $t' \in T(C) \cap S^{*\circ}(x + y')$. Then $t - t' \in T(C') \cap S^{*\circ}(y - y') \subseteq T_1(C')$ and as $t = t' + (t - t')$, our claim is proved.

Using quantifier elimination in $(J, S, S^{*\circ})$, the claim gives

$$tp(a/\bar{a}, T(C)) \cup tp(a/\bar{a}, T_1(C')) \implies tp(a/\bar{a}, T(C')).$$

Now in order to show

$$tp(a/C') \cup gtp(a/\,acl(C') \cap A^*) \implies^* tp(a/C')$$

it will suffice to check that

$$(**)\qquad tp(a/\bar{a}) \cup gtp(a/C' \cap A^*) \implies tp(a/\bar{a}, T_1(C')).$$

We fix C' and let $\pi : L \to \bar{A} \times A_1^*$ be the semi-dual cover corresponding to $(0) \to A_1 \to A \to \bar{A} \to (0)$. Let D' be the locus of $\bar{a}$ over C'. If $t \in T_1(C')$, then $(\bar{a}, t) \in L$; let $a^* = \pi_2(\bar{a}, t)$ be the induced element of A^*. Then $a^* \in C' \cap A_1^*$. As $t \in dcl(\bar{a}, C')$ we may write $(\bar{a}, t) = g(\bar{a}) = (\bar{a}, g_2(\bar{a}))$, where $g : D' \to L(a^*)$ is a C'-definable section. By Lemma 6.6.8 there is a C'-definable homomorphism j onto a finite group whose values determine $g_2(\bar{u})(u)$ for $u \in A$, $\bar{u} \in D'$. By definition $gtp(a/C' \cap A^*)$ determines the value of $j(a)$ and hence of $t(a)$. Claim $(**)$ follows.   ∎

Proof of Proposition 6.6.1. We proceed by induction on the length of a maximal chain of $acl(\emptyset)$-definable subgroups. We may work over $acl(\emptyset)$. If A contains a finite subgroup defined over $acl(\emptyset)$ we may apply induction and Lemma 6.6.6. Accordingly we may suppose $acl(\emptyset) \cap A = (0)$. Similarly we may suppose $acl(\emptyset) \cap A^* = (0)$, using Lemma 6.6.5, part (2).

Now A contains an $acl(\emptyset)$-definable rank 1 subgroup A_1 which is part of a basic linear geometry J (Lemmas 6.2.6, 6.2.11). If A_1 has an $acl(\emptyset)$-definable complement A_2 then we may assume both A_1 and A_2 are settled, and then $A = A_1 \oplus A_2$ is settled. Accordingly we may suppose that A_1 is not complemented. Now by induction A/A_1 is settled over some set C and after enlarging C if necessary, we may assume that the associated linear geometry is not quadratic (adding an element of the quadratic space Q, if needed). Now the previous lemma applies.   ∎

The following is another version of the finite basis property.

Proposition 6.6.11. *Let $\mathcal{M}$ be Lie coordinatizable and A an abelian group interpreted in $\mathcal{M}$. Then there is a finite collection D_i of definable subsets of A, such that every definable subset of A is a boolean combination of translates of the D_i together with cosets of definable subgroups.*

Proof. We proceed by induction on $rk(A)$. Let D_i be a finite list of definable sets including all the definable sets associated correspondingly to all $acl(\emptyset)$-definable subgroups of smaller rank. In addition let C be a finite set over which A is settled, and assume that all C-definable sets occur as well in the list (D_i). We claim this suffices.

As A is settled over C, it will suffice to consider definable subsets D of A of rank less than $rk\, A$. Such a set lies in the union of a finite number of cosets of $acl(\emptyset)$-definable subgroups of A of rank less than $rk(A)$, by Lemma 6.2.5 and Proposition 6.2.3. We may therefore assume that D lies in one such coset, and since our problem is invariant under translation, we may even assume D lies in an $acl(\emptyset)$-definable subgroup of smaller rank, and conclude.   ∎

7

Reducts

7.1 RECOGNIZING GEOMETRIES

Our main objective in the present section is to characterize coordinatizing geometries as follows.

Proposition 7.1.1. *Let $\mathcal{M}$ be $\aleph_0$-categorical of finite rank, and let A, A^* be rank 1 groups equipped with vector space structures over a finite field F, and a definable F-bilinear pairing into F, with everything 0-definably interpreted in $\mathcal{M}$. Assume the following properties:*

L1. *Every $\mathcal{M}$-definable F-linear map $A \to F$ is represented by some element of A^*, and dually.*

L2. *Algebraic closure and linear dependence coincide on A and on A^*.*

L3. *A and A^* have no nontrivial proper 0-definable subspaces.*

L4. *Every definable subset of A or of A^* is a boolean combination of 0-definable subsets and cosets of definable subgroups.*

L5. *If D is the locus of a complete type in A over $acl(\emptyset)$, and $a'_1, \ldots, a'_n$ are F-linearly independent elements of A^*, then there is an element d of D with (d, a'_i) prescribed arbitrarily.*

Then the pair (A, A^*) is a linear Lie geometry, possibly weak, which is stably embedded in $\mathcal{M}$.

The proof will require a number of preliminary lemmas. We remark that in view of hypothesis (L3), either one of the groups A, A^* vanishes (in which case we might as well assume $A^* = (0)$), or the pairing is nondegenerate on both sides. In the latter case the notation A^* is justified by hypothesis (L1).

We will continue to label the various hypotheses as in the statement of Proposition 7.1.1.

Lemma 7.1.2. *Let $\mathcal{M}$ be $\aleph_0$-categorical of finite rank and let A, A^* be rank 1 groups equipped with vector space structures over a finite field F, and a definable F-bilinear pairing into F, with everything 0-definably interpreted in $\mathcal{M}$. Assume that*

L2. *Algebraic closure and linear dependence coincide on A and on A^*.*

L3. *A and A^* have no nontrivial proper 0-definable subspaces.*

Then either A and A^* are algebraically independent, or there is a 0-definable bijection between their projectivizations P and P^*.

Proof. This is the standard nonorthogonality result. We assume an algebraic relation between A and A^*, specifically $rk(\mathbf{a}) = k$, $rk(\mathbf{a}^*) = k^*$, $rk(\mathbf{aa}^*) < k + k^*$ with $\mathbf{a} \in A$ and $\mathbf{a}^* \in A^*$. We will first find an element of A algebraic over $\mathbf{a}^*$. Suppose $\mathbf{a}$ is not itself algebraic over $\mathbf{a}^*$. Then we take independent conjugates $\mathbf{a}_i$ of $\mathbf{a}$ over $acl(\mathbf{a}^*)$ and find $rk(\mathbf{a}_1, \ldots, \mathbf{a}_n) < nk$ for n large. By the dimension law in projective space there is then $a \in A - (0)$ in $acl(\mathbf{a}_1, \ldots, \mathbf{a}_i) \cap acl(\mathbf{a}_{i+1}, \ldots, \mathbf{a}_n)$ and hence algebraic over $\mathbf{a}^*$.

Switching sides, we may then find $a^* \in A^* - (0)$ algebraic over a. Then $acl(a) = acl(a^*)$ and this gives a bijection between a subset of P and a subset of P^*. Furthermore, the argument shows that the domain and range of the bijection are algebraically closed, and thus correspond to 0-definable subspaces of A and A^*. By hypothesis (L3) the bijection is total. ∎

Lemma 7.1.3. *Let $\mathcal{M}$ be $\aleph_0$-categorical of finite rank and let A, A^* be rank 1 groups equipped with vector space structures over a finite field F, and a definable F-bilinear pairing into F, with everything 0-definably interpreted in $\mathcal{M}$. Assume that*

L1. *Every $\mathcal{M}$-definable F-linear map $A \to F$ is represented by some element of A^*, and dually.*

L2. *Algebraic closure and linear dependence coincide on A and on A^*.*

Assume in addition that the projectivizations P, P^* of A and A^* correspond by a 0-definable bijection. Then there is an identification of A with A^* according to which the given pairing $A \times A^* \to F$ is symplectic, unitary, or orthogonal.

Proof. As P and P^* are definably isomorphic, there is a semilinear isomorphism of A with A^*, which gives rise to a self-pairing $A \times A \to F$ which is linear in the first variable and satisfies $(x, \alpha y) = \alpha^\sigma (x, y)$ with an automorphism σ on the right. In particular the map $\lambda_x : A \to A$ defined by $(x, y)^{\sigma^{-1}}$ is F-linear and hence by hypothesis is given by a unique element x^*: $(y, x^*)^\sigma = (x, y)$. As x^* is definable from x, we have $x^* = \alpha x$ for some $\alpha = \alpha(x) \in F$ possibly dependent on x.

We have

$$(y, (\beta x)^*)^\sigma = (\beta x, y) = \beta(x, y) = \beta(y, x^*)^\sigma = (\beta^{\sigma^{-1}} (y, x^*))^\sigma$$
$$= (y, \beta^{\sigma^{-2}} x^*)^\sigma$$

and thus $(\beta x)^* = \beta^{\sigma^{-2}} x^*$. Now for x_1, x_2 linearly independent with $\alpha(x_1) = \alpha(x_2) = \alpha_\circ$ we have $(x_1 + \beta x_2)^* = \alpha_\circ(x_1 + \beta^{\sigma^{-2}} x_2)$, and as the latter is a scalar multiple of $x_1 + \beta x_2$, we find that σ^2 is the identity and x^* is a linear function of x. The same computation shows that for x_1, x_2 linearly independent, $\alpha(x_1) = \alpha(x_2)$, and thus $\alpha(x)$ is independent of x; so $x^* = \alpha x$ for a fixed α:

$$(x, y) = \alpha(y, x)^\sigma.$$

Applying this law twice, $(x, y) = \alpha \alpha^\sigma (x, y)$ and

$$\alpha \alpha^\sigma = 1.$$

If σ is the identity, then $\alpha = \pm 1$ and the form (x, y) is either symmetric or symplectic. In characteristic 2 we conclude only that it is symmetric, but in this case the form (x, x) is the square of a linear functional and vanishes on a subspace of codimension at most 1. If we exclude 0-definable proper subspaces of finite codimension we may conclude that in characteristic 2 the form is symplectic.

When σ is nontrivial we have in any case the norm of α equal to 1 and thus $\alpha = \gamma^\sigma / \gamma$ for some $\gamma \in F$. Then one checks that $\gamma(x, y)$ is a unitary form on A. ∎

Definition 7.1.4. *The geometric language for (A, A^*) consists of the F-space structure, the pairing, an identification of A with A^* as above, if available, and all $acl(\emptyset)$-definable subsets of A and A^*. Vector space operations and the identification, if present, are taken as functions, rather than being encoded by relations.*

We are working over $acl(\emptyset)$ here. The identification between A and A^* depends in the unitary case on a parameter from the fixed field of the automorphism, but is algebraic over $acl(\emptyset)$.

Lemma 7.1.5. *Let $\mathcal{M}$ be $\aleph_0$-categorical of finite rank and let A, A^* be rank 1 groups equipped with vector space structures over a finite field F, and a definable F-bilinear pairing into F, with everything 0-definably interpreted in $\mathcal{M}$. Assume*

L1. *Every $\mathcal{M}$-definable F-linear map $A \to F$ is represented by some element of A^*, and dually.*
L2. *Algebraic closure and linear dependence coincide on A and on A^*.*
L3. *A and A^* have no nontrivial proper 0-definable subspaces.*
L4. *Every definable subset of A or of A^* is a boolean combination of 0-definable subsets and cosets of definable subgroups.*
L5. *If D is the locus of a complete type in A over $acl(\emptyset)$ and $a'_1, \ldots, a'_n$ are F-linearly independent, then there is an element d of D with (d, a'_i) prescribed arbitrarily.*

Then the induced structure on (A, A^*) admits quantifier elimination in the geometric language.

Proof. This may seem obvious; but condition (L4) is rather vague as to the provenance of the parameters involved.

We show by induction on n that the quantifier-free type of $a_1, \ldots, a_n$ determines its full type. If A and A^* are identified, we work in A exclusively. By hypothesis (L2) we may suppose the a_i are algebraically independent.

We will establish the following for any finite set C and any C-definable subset D of A:

$(*)$ D is a boolean combination of 0-definable sets, a finite subset of $acl(C)$, and cosets of the form $H_\alpha = \{x \in A : (x, c) = \alpha\}$ with $c \in A^*$ algebraic over C.

Assuming the claim, let C be $acl(a_1, \ldots, a_{n-1}) = dcl(a_1, \ldots, a_{n-1})$. By our induction hypothesis the type of C is known. By $(*)$ the type $tp(a_n/C)$ is determined by its atomic type over C, and hence over $a_1, \ldots, a_{n-1}$, since C is generated by functions over $a_1, \ldots, a_{n-1}$.

It remains to establish $(*)$. We may suppose that the set D is the locus of a complete nonalgebraic type over $acl(C) = dcl(C)$. Let D' be the minimal $acl(\emptyset)$-definable set containing D. We note first that in hypothesis (L4) we may take the definable subgroups involved to be subspaces of finite codimension. Indeed, if B is an infinite definable subgroup of A then it has finite index in A and the intersection of αB for $\alpha \in F^\times$ is a definable subspace of finite codimension contained in B. Thus modulo the ideal of finite sets, D is the intersection with D' of a boolean combination D_1 of translates of definable subspaces of finite codimension. There is a definable linear map θ from A to a finite dimensional space F^n, and a subset X of F^n, such that $D_1 = \theta^{-1}[X]$. Minimize n. We may represent θ as $(a_1^*, \ldots, a_n^*)$ for some $a_i^* \in A^*$. We claim the a_i^* lie in $acl(C)$. We may in any case assume $a_i \in acl(C)$ for $i \leq n_0$ and the remaining a_i are algebraically independent over $acl(C)$. If $n_0 < n$ then let $a'_{n_0+1}, \ldots, a'_n$ be conjugate to $a^*_{n_0+1}, \ldots, a^*_n$ over C and linearly independent from $a_1^*, \ldots, a_n^*$. As n has been minimized we can find $\alpha \in F^{n_0}$ and $\beta, \beta' \in F^{n-n_0}$ with $(\alpha, \beta) \in X$, $(\alpha, \beta') \notin X$. Applying (L5), we may find infinitely many elements $d \in D'$ satisfying

$$(d, a_i^*) = \alpha_i; (d, a^*_{n_0+i}) = \beta_i; (d, a'_{n_0+i}) = \beta'_i.$$

Off a finite set this yields $d \in D$ and $d \notin D$, a contradiction. Thus the a_i^* are algebraic over C. Finally, the finite set involved is the difference of two sets defined over $acl(C)$ and hence lies in $acl(C)$. ∎

Proof of Proposition 7.1.1. In view of Lemma 7.1.5, to complete the analysis of (A, A^*), we must determine the 0-definable subsets of A (and similarly, A^*) more or less explicitly. Let P be the set of types of nonzero elements of A over $acl(\emptyset)$. For $a \in A$ set $q(a) = tp(a/acl(\emptyset))$. Note that these types have rank

1, with the exception of $tp(0/acl(\emptyset))$. By the proof of the previous lemma, if a and b are algebraically independent elements of A then the type of $a + b$ over $acl(\emptyset)$ is determined by: $q(a)$, $q(b)$, and $(a, b) \in F$. (When there is no identification of A with A^*, let the form (a, b) be identically 0 on A.) Thus $q(a + b) = f(q(a), q(b), (a, b))$ for some function $f : P \times P \times F \to P$.

Consider $+ : P^2 \to P$ defined by $p_1 + p_2 = f(p_1, p_2, 0)$. We claim that $+$ is an abelian group operation on P. This operation is clearly commutative. For associativity, let $p_1, p_2, p_3 \in P$. We may assume they are all nonzero. By type amalgamation and the hypothesis (L5) we can find a_1, a_2, a_3 independent with the prescribed types and with $(a_i, a_j) = 0$ for distinct i, j. Then $p_1 + p_2 + p_3$, computed in either possible way, will give $q(a + b + c)$. Finally we check cancellation. Suppose $p_\circ + p_1 = p_\circ + p_2$. We may then choose independent $a_\circ, a_1, a_2$ realizing the prescribed types, with $(a_\circ, a_1) = (a_\circ, a_2) = 0$, and we find that $q(a_\circ + a_1) = q(a_\circ + a_2)$ and $(-a_\circ, a_\circ + a_1) = (-a_\circ, a_\circ + a_2) = -(a, a)$. Thus $q(a_1) = f(q(-a), p_\circ + p_1, -(a, a)) = q(a_2)$, as claimed.

Thus P is a finite abelian group. Let the zero element of P be denoted $p_\circ$, and let D be the locus of this type in A.

We now dispose of the polar case, in which there is no identification of A with A^*. Then $q : A \to P$ is generically a homomorphism and hence extends to a homomorphism by sending 0 to 0. As A has no proper 0-definable subspace of finite codimension, it has no proper 0-definable subgroup of finite index, and thus the homomorphism is trivial, and $A - (0)$ realizes a unique type over $acl(\emptyset)$. This completes the analysis of the polar case.

For the remainder of the argument we may suppose that A and A^* have been identified, or, in other words, that A carries a symmetric, symplectic, or unitary form. If P consists of a single type, then this form is symplectic and the types are entirely known. We may assume therefore that P contains more than one type. It is of course still possible that the form is symplectic.

D is infinite, and is the locus of a type over $acl(\emptyset)$, and hence generates A. The group $Stab(D)$ has rank 1, and hence coincides with A. Thus a generic element of A belongs to $Stab_\circ(D)$ and can therefore be expressed as $a + b$, with $a, b \in D$ independent. As the type of $a + b$ is determined by the value of (a, b), for $a, b \in D$ independent, this gives rise to a function $f^* : F \to P$.

For independent $a, b, c \in D$ with $(a, b) = 0$ we have

$$q(a + b) = q(a) + q(b) = p_\circ$$

and thus $a + b \in D$, and as $(a + b, c) = (a, c) + (b, c)$ it follows that f^* is an additive homomorphism. We define a map $\nu : F^\times \to End(P)$ by $\nu(\alpha) \cdot q(a) = q(\alpha a)$. This is clearly a well-defined multiplicative homomorphism into $End(P)$. In particular $p_\circ$ is fixed by $\nu[F^\times]$, and thus D is invariant under nonzero scalar multiplication. Thus we may make the following computation with $a, b \in D$ independent, $(a, b) = \alpha$:

$$(*) \qquad f^*(\beta\beta^\sigma\alpha) = q(\beta a + \beta b) = \nu(\beta)q(a + b) = \nu(\beta)f^*(\alpha).$$

Now let K be the kernel of f^*, and $F_\circ$ the fixed field of σ (which may be all of F). We will show that $K = \ker Tr$ with Tr the trace from F to $F_\circ$, which will allow us to identify P and $F_\circ$.

By $(*)$ K is invariant under multiplication by elements $\beta\beta^\sigma$, that is by norms or squares according as σ is nontrivial or trivial, and therefore is an $F_\circ$-subspace of F in all cases. Furthermore, $K < F$ since P has more than one element. Thus if σ is the identity and $F_\circ = F$ we have only the possibility $K = (0)$, which is the claim in this case. Suppose now that σ is nontrivial, so that F is a quadratic extension of $F_\circ$. As $q(x + y) = q(y + x)$ we get $f^*(\alpha) = f^*(\alpha^\sigma)$ so K contains the kernel $\{\alpha - \alpha^\sigma : \alpha \in F\}$ of the trace, which is of codimension 1 in F. Thus K coincides with this kernel.

Accordingly, we now identify P with $F_\circ$ and f^* with the trace. The formula $(*)$ then states that ν is the norm if σ is nontrivial, and the squaring map otherwise. In particular there are $|F_\circ|$ nontrivial types over $acl(\emptyset)$. These types must therefore be determined by the function (x, x), unless the form is symplectic.

Suppose, finally, that the form is symplectic; we still suppose that $|P| = |F_\circ|$. Take x, y independent and orthogonal. Then $(x - y, y) = 0$ and thus $q(x) = q(x-y)+q(y) = q(x)+q(-y)+q(y)$, that is $q(-y) = -q(y)$. On the other hand, by $(*)$ we have $q(-y) = q(y)$, and thus the characteristic is 2. Our final objective is to show that q is a quadratic form, so that A is an orthogonal space in characteristic 2. In any case, $(*)$ says that $q(\alpha x) = \alpha^2 q(x)$, and it remains to study $q(x + y)$.

Take x_1, x_2, y_1, y_2 in D independent with x_i orthogonal to y_i for $i = 1, 2$, and let $\alpha = (x_1, x_2)$, $\beta = (y_1, y_2)$. Let $z_i = x_i + y_i$; then $z_i \in D$ and

$$q(z_1 + z_2) = (z_1, z_2) = \alpha + \beta + (x_1, y_2) + (x_2, y_1).$$

Let $x = x_1 + x_2$ and $y = y_1 + y_2$. Then x and y are independent; $q(x) = \alpha$ and $q(y) = \beta$; and $(x, y) = (x_1, y_2) + (x_2, y_1)$. As $x + y = z_1 + z_2$, we have $q(x+y) = q(x)+q(y)+(x, y)$. This argument applies to x, y independent and nonzero. When x, y are dependent they are linearly dependent, and it follows easily that this formula holds in general. Thus q is a quadratic form associated to the given symplectic form. This determines the structure of A in this last case.   ∎

Lemma 7.1.6. *Let $\mathcal{M}$ be $\aleph_0$-categorical of finite rank. Let A, A^* be 0-definably interpreted rank 1 vector spaces over a finite field F with a definable F-bilinear pairing satisfying*

L1. *Every $\mathcal{M}$-definable F-linear map $A \to F$ is represented by some element of A^*, and dually.*

L2. *Algebraic closure and linear dependence coincide on A and on A^*.*

L3. *A and A^* have no nontrivial proper 0-definable subspaces.*

Suppose that over $acl(\emptyset)$, A, A^* are part of a linear Lie geometry sta-

bly embedded in $\mathcal{M}$. *Then* A, A^* *are part of a linear Lie geometry stably embedded in* $\mathcal{M}$.

Proof. We have to show that if A carries a bilinear form or quadratic form defined over $acl(\emptyset)$ then the set of scalar multiples of the form is 0-definable, and similarly if A, A^* are part of a quadratic geometry in characteristic 2.

Note that any $acl(\emptyset)$-definable linear automorphism of A acts trivially on the projective space PA, by (L2), and hence is given by a scalar multiplication. As A^* contains all definable linear forms on A, any two nondegenerate bilinear forms differ by a definable automorphism of A, hence differ by a scalar. In odd characteristic this disposes of all cases since quadratic forms correspond to inner products.

Consider now the case of a symplectic space in characteristic 2, where the form is known up to a scalar multiple. With the form fixed, the set of quadratic forms compatible with it and definable over $acl(\emptyset)$ corresponds to $A^* \cap acl(\emptyset)$. By (L3) this is (0). Thus if there are quadratic forms definable over $acl(\emptyset)$, they are the scalar multiples of a single form.

Suppose, finally, that there are no $acl(\emptyset)$-definable quadratic forms but that there is an $acl(\emptyset)$-definable quadratic geometry. In this case, the set of $acl(\emptyset)$-definable quadratic forms compatible with one of the bilinear forms carries a regular action by A^*; hence this is the standard quadratic geometry over $\emptyset$, corresponding to a form known up to a scalar multiple. Note that the pairing is known but the identification of A with A^* is known only up to a scalar multiple. ∎

Proposition 7.1.7. *Let* $\mathcal{M}$ *be* $\aleph_0$-*categorical of finite rank. Let* A, A^* *be* 0-*definably interpreted rank 1 vector spaces over a finite field* F *with a definable* F-*bilinear pairing satisfying*

L1. *Every* $\mathcal{M}$-*definable* F-*linear map* $A \to F$ *is represented by some element of* A^*, *and dually.*

L3. A *and* A^* *have no nontrivial proper* 0-*definable subspaces.*

Let $c \in \mathcal{M}$, *with* $acl(c) \cap (A, A^*) = dcl(c) \cap (A, A^*)$ *nondegenerate, and set* $(A', A'^*) = [acl(c) \cap (A, A^*)]^{\perp}$. *Assume that relative to a possibly larger field* F', *in* $\mathcal{M}' = \mathcal{M}$ *with* c *added as a constant,* (L1,L3) *hold for* A', A'^* *as well as:*

L2 *Algebraic closure (over* c *and linear dependence (over the extended scalar field) coincide on* A' *and on* A'^*.

L4' *Every definable subset of* A' *or of* A'^* *is a boolean combination of* c-*definable subsets and cosets of definable subgroups.*

L5' *If* D *is the locus of a complete type in* A' *over* $acl(c)$ *and* $a'_1, \ldots, a'_n$ *are* F-*linearly independent, then there is an element* d *of* D *with* (d, a'_i) *prescribed arbitrarily.*

Then there is a 0-definable sort Q in $\mathcal{M}$ such that (A, A^*, Q) form a weak linear Lie geometry, stably embedded in $\mathcal{M}$.

Proof. We will work over $acl(\emptyset)$. We let Q be $\emptyset$ unless A carries an $acl(\emptyset)$-definable symplectic bilinear form in characteristic 2, in which case we let Q be the set of all definable quadratic forms which are compatible with one of these symplectic forms on A; each component of this set, corresponding to a particular form, has a regular action by A^* and is, in particular, uniformly definable. Thus Q is 0-definable. We let $J = (A, A^*, Q)$, equipped with all structure defined over $acl(\emptyset)$, and we claim that this is stably embedded.

Let $\mathcal{M}'$ be the expansion of $\mathcal{M}$ by the constant c, and J' the geometry A', A'^* with the structure inherited from $\mathcal{M}'$. By Proposition 7.1.1, J' is a stably embedded weak linear geometry. Let $A_\circ = acl(c) \cap A$. Then $A = A_\circ \oplus A'$, and similarly for A^*, and Q. Thus J is contained in the definable closure of J' in $\mathcal{M}'$. Thus J inherits the following properties:

> J is stably embedded in $\mathcal{M}$;
> J has finite rank and is modular;
> J has the type amalgamation property of Proposition 5.1.15.

By Proposition 6.2.3, if H is a parametrically definable subgroup of $A \times A$ or $A \times A^*$ in $\mathcal{M}$, then H is commensurable with an $acl(\emptyset)$-definable subgroup.

Let F' be the ring of endomorphisms of A which are 0-definable in J. By the third hypothesis, F' is a field, and it must restrict to a subfield of the field of scalars for J'. We claim, in fact, that F' induces the scalars of J'. Let α be one of the scalar multiplications on J'. The graph of α is commensurable with an $acl(\emptyset)$-definable subgroup H of $A \times A$. By the third condition, H is the graph of a group isomorphism from A to A. Let $\alpha \in F'$ be the element with graph H. As the graphs of α and α' are commensurable $acl(c)$-definable automorphisms of A', they agree there.

The same sort of argument shows that an isomorphism $A' \to A'^*$ is induced by an $acl(\emptyset)$-definable isomorphism on A of the same type. The same applies to quadratic forms in odd characteristic since they correspond to bilinear forms. In characteristic 2 one can, in any case, extend quadratic forms to forms on A in $acl(c)$, taking them to vanish on $acl(c) \cap A$.

Now let J^- be J reduced to its geometric structure. The structure on J' is known and is defined from this geometric structure by Proposition 7.1.1. As J is interpreted in J', every 0-definable relation in J is definable in J^- from parameters in $acl(c)$. Let R be 0-definable in J, with canonical parameter $e \in J^-$, and definable in J^- from the parameter a. By weak elimination of imaginaries in J^- we may take $a \in acl(e)$ in J^-; but $e \in acl(\emptyset)$ in J, so $a \in acl_J(\emptyset) \cap J^-$, which is trivial by assumption. Thus R is 0-definable in J^- and $J = J^-$ is a stably embedded Lie geometry.

This argument took place over $acl(\emptyset)$ (and our last 0-definability claim is blatantly false in general); to remove this, we use the preceding lemma. ∎

Remark 7.1.8. We are dealing in Proposition 7.1.1 with the rank 1 case of the analysis of settled groups with $acl(\emptyset) \cap A = (0)$, $acl(\emptyset) \cap A^* = (0)$. It would be interesting to tackle the general case. Two special cases: analyze the case of prime exponent, or the case of rank 2.

7.2 FORGETTING CONSTANTS

The following is a special case of Proposition 7.5.4 below, for which we will give a proof by a method not depending on the classification of finite simple groups. The proof given here goes via smooth approximation rather than coordinatization and involves [KLM], hence the classification of the finite simple groups.

Proposition 7.2.1. *Let M be a structure and M_c an expansion of M by a constant c. If M_c is smoothly approximable by finite structures, then there is an expansion M° of M by an algebraic constant which is smoothly approximable.*

The key example here is due to David Evans: one takes M to be the reduct of a basic quadratic geometry in which the orientation is forgotten, but the corresponding equivalence relation is remembered. In a finite approximation the two classes are distinguished, so M is not smoothly approximable by finite structures. The orientation itself is an algebraic constant. It can be shown that this is the only sort of algebraic constant which comes in to Proposition 7.2.1.

Definition 7.2.2. *If M is Lie coordinatizable and E is an envelope in M it is said to be* equidimensional *if all the isomorphism types of specified geometries of a given type are the same; that is, the dimensions and Witt defects are constant.*

Lemma 7.2.3. *Let N be smoothly approximable, $c \in N$, E a finite subset of N containing c. Then*

1. *If E is an envelope of N, it is an envelope of N_c.*
2. *If E is an equidimensional envelope of N_c, it is an envelope of N, provided that:*
   (i) *The locus of c over $\emptyset$ is nonmultidimensional;*
   (ii) *For any $acl(\emptyset)$-definable canonical projective geometry P_b, with b its canonical parameter, we have $tp(b)$ implies $tp(b/c)$.*

Proof. We use the criterion given in the corollary to Lemma 3.2.4. Of the three conditions given there, only the last one is actually sensitive to the presence of the parameter c. In $\mathcal{N}$ this may be phrased as follows:

> If c_1, c_2 are conjugate in $\mathcal{M}$ and D_{c_1}, D_{c_2} are corresponding conjugate definable sets, then $D_{c_1} \cap E$ and $D_{c_2} \cap E$ are conjugate by an elementary automorphism of E.

This condition is certainly inherited "upward," giving the first point. For the second, assuming conditions (i) and (ii), and the conjugacy condition in $\mathcal{N}_c$, it suffices to to show the conjugacy condition for canonical projective geometries D_{c_i}. There are two cases.

Suppose first that $c_i \notin acl(\emptyset)$. Then D_{c_i} is orthogonal to $tp(c/c_i)$ as the latter is analyzed by $acl(\emptyset)$-definable geometries. Hence D_{c_i} remains a projective geometry in $\mathcal{N}_c$. It is also canonical: every proper conjugate in $\mathcal{N}_c$ is in particular a conjugate in $\mathcal{N}$, and hence orthogonal to D_{c_i}. Thus the dimension of D_{c_i} in E is one of the specified dimensions as an envelope in $\mathcal{N}_c$; these are all assumed equal, so D_{c_1} and D_{c_2} have the same dimension and similarly, where applicable, the same Witt defect.

Now suppose $c_i \in acl(\emptyset)$. Then by 2(ii) $tp(c_1/c) = tp(c_2/c)$ and thus they are conjugate in E_c, and the $D_{c_i} \cap E$ are conjugate.   ∎

We now deal with a special case of Proposition 7.2.1.

Lemma 7.2.4. *Let $\mathcal{M}$ be a structure and $\mathcal{M}_c$ an expansion of $\mathcal{M}$ by a constant c. Assume that the locus P of c in $\mathcal{M}$ is nonmultidimensional in $\mathcal{M}_c$ and that for any $acl(c)$-definable canonical projective geometry J_b, $tp(b/c)$ implies $tp(b/acl(c))$. If $\mathcal{M}_c$ is smoothly approximable by finite structures, then there is an expansion $\mathcal{M}^\circ$ of $\mathcal{M}$ by an algebraic constant which is smoothly approximable.*

Proof. An envelope in $\mathcal{M}_c$ is determined by a k-tuple of dimensions for some k. Let q be a 2-type realized in P. Define a binary relation R_q between k-tuples of dimensions as follows: $R_q(d, d')$ if and only if there is a realization (c, c') of q, and a finite subset E of $\mathcal{M}$ which is an envelope of dimension d in $\mathcal{M}_c$ and is an envelope of dimension d' in $\mathcal{M}_{c'}$. We claim that R_q defines a partial function. If (c, c') realizes q, then $tp(c'/c)$ in $\mathcal{M}_c$ determines $tp(c'/c)$ in U and hence determines the corresponding dimension d'. We will use function notation, writing $f_q(d) = d'$.

We define an equivalence relation on P as follows: $E(a, b)$ holds if there is a finite subset C_o of P such that for any finite subset C of P containing C_o, any equidimensional envelope of $\mathcal{M}_C$ is an envelope of $\mathcal{M}_a$ and $\mathcal{M}_b$, with the same dimensions. We claim:

> If $a, b, b' \in P$ and $tp(ab) = tp(ab')$, then $E(b, b')$.

Given such a, b, b' we let $q = tp(ab) = tp(ab')$ and $C_o = \{a, b, b'\}$. If C contains a, b, b' and U is an equidimensional envelope of $\mathcal{M}_C$, then U is an equidimensional envelope over a, b, or b'; and the dimension over b or b' is f_q applied to the dimension over a.

Thus the relation E has finitely many equivalence classes. Let c_o be the class $c/E \in acl(\emptyset)$. We claim that $\mathcal{M}$ is smoothly approximable over c_o.

Let P be the increasing union of finite subsets C_n with $C_1 = \{c\}$ and let U_n be an n-equidimensional envelope in $\mathcal{M}_{C_n}$ containing U_{n-1}. Let $\mathcal{M}$ be the canonical language for $\mathcal{M}$ (consisting of complete types over $\emptyset$). Let $\mathcal{F}$ be a nonprincipal ultrafilter on ω and let the term "almost all n" be understood with reference to this ultrafilter. Let $\mathcal{M}^*$ be the set of relations which are 0-definable in $\mathcal{M}(c)$ whose restrictions to U_n are L-definable for almost all n. We will show that $\mathcal{M}^* = L(c_o)$ and that $\mathcal{M}$ is smoothly approximable in the language L^*.

$\mathcal{M}^*$ is a sublanguage of $\mathcal{M}(c)$ which contains $\mathcal{M}(c_o)$ since the proof that E has finitely many classes also shows c_o is definable in U_n from some point on. To see that $\mathcal{M}$ is smoothly approximable in the language $\mathcal{M}^*$, let k be fixed and let $\mathbf{a}, \mathbf{b}$ be k-tuples with the same type in $\mathcal{M}^*$. It suffices to show that for almost all n, two such k-tuples in U_n will be conjugate in U_n. If not, then for almost all n, there is a 0-definable k-ary relation R_n on U which does agree on U_n with any relation in $\mathcal{M}^*$. However, it must agree with some c-definable relation restricted to U_n, and there are only finitely many such, so for almost all n R_n agrees with the same c-definable relation on U_n, which means it agrees with a relation of $\mathcal{M}^*$, a contradiction.

It remains to be shown that $\mathcal{M}^* \subseteq \mathcal{M}(c_o)$. Let P' be the equivalence class of c with respect to E; this is a subset of P. We claim first that

$$P' \text{ realizes a unique } \mathcal{M}^*\text{-type.}$$

Take $c' \in P'$. It suffices to show that for almost all n, and indeed for all sufficiently large n, there is an automorphism of U_n carrying c to c'. For large n, U_n contains c and c' and is an equidimensional envelope with the same dimensions relative to c and to c'. Thus $\mathcal{M}_c$ and $\mathcal{M}_{c'}$ are isomorphic smoothly approximable models and U_n over c or c' is an equidimensional envelope with respect to the same data in both cases; by uniqueness of envelopes, $(U_n, c) \simeq (U_n, c')$.

It follows that any automorphism σ of $\mathcal{M}_{c_o}$ preserves $\mathcal{M}^*$: as σ preserves P', by the previous claim we may suppose that σ fixes c, and hence $\mathcal{M}^*$. Thus $\mathcal{M}^* \subseteq \mathcal{M}(c_o)$. ∎

Lemma 7.2.5. *Let $\mathcal{M}$ be smoothly approximable, and for $a \in \mathcal{M}$ let $a^{(1)} = \{a' \in acl(a) : rk(a') = 1\}$. Define $E(a, b)$ by: $a^{(1)} = b^{(1)}$. Then we have:*

1. *If S is an $acl(\emptyset)$-definable subset of $\mathcal{M}$ of rank $n > 0$, then each E-class in S has rank less than n.*
2. *$\mathcal{M}/E$ is nonmultidimensional.*

3. *If $c \in \mathcal{M}$ and a and b are both independent from c, then $a^{(1)} = b^{(1)}$ if and only if the same relation holds in $\mathcal{M}_c$.*

Proof. The first point is the coordinatization theorem, i.e., without loss of generality $\mathcal{M}$ is Lie coordinatized. The second point is clear as the 0-definable closure of $a^{(1)}$ is a set of rank at most 1 over $\emptyset$.

For the final point, write $a_c^{(1)}$ for $a^{(1)}$ computed over c. We wish to show that each of $a^{(1)}$, $a_c^{(1)}$ determines the other. As

$$a^{(1)} = \{a' \in a_c^{(1)} : a' \text{ is independent from } c\}$$

it suffices to deal with the reverse direction. We claim that

$$a_c^{(1)} = acl(a^{(1)}, c).$$

In any case, the right side is contained in the left. Conversely, we must show that if $d \in acl(a, c)$ has rank at most 1 over c then $d \in acl(a^{(1)}, c)$. By modularity a and c, d are independent over $a' = acl(a) \cap acl(c, d)$. Thus a and d are independent over $a'c$ and therefore $d \in acl(a'c)$. But $rk(a'/c) \leq 1$ and a, c are independent, so $rk(a') \leq 1$. Thus $a' \in a^{(1)}$ and $d \in acl(a^{(1)}, c)$.   ∎

Proof of Proposition 7.2.1. We assume $\mathcal{M}_c$ is smoothly approximable and we seek $c_o \in acl(\emptyset)$ with $\mathcal{M}_{c_o}$ smoothly approximable. We work over $acl(\emptyset)$, and we replace c by a finite subset C of $acl(c)$ such that for P_b an $acl(c)$-definable canonical projective geometry, $tp(b/C)$ implies $tp(b/acl(c))$. We again write c rather than C. After these adjustments, if the locus P of c is nonmultidimensional, then Lemma 7.2.4 applies. We treat the general case by induction on $rk\,c$.

If there is $c_1 \in acl(c)$ with $c \notin acl(c_1)$, then after expanding c_1 if necessary to a slightly larger subset of $acl(c_1)$ we may take $\mathcal{M}_{c_1}$ to be smoothly approximable, by induction, as $rk(c/c_1) < rk(c)$, and then by a second application of induction, as $rk(c_1) < rk(c)$, we reduce to a parameter in $acl(\emptyset)$. We assume therefore that there is no such element c_1.

We define a relation E on P as follows: $E(a, b)$ holds if for some $c \in P$ independent from a, b we have $a_c^{(1)} = b_c^{(1)}$; here $a_c^{(1)}$ is $a^{(1)}$ computed over c, as in the previous lemma. We claim that if $c, c' \in P$ are both independent from ab and $a_c^{(1)} = a_{c'}^{(1)}$, then the same applies over c'. Working with an element c'' independent from a, b, c, c', we reduce to the case in which c and c' are independent over a, b; in other words, the triple ab, c, c' is independent. As $\mathcal{M}_c$ is smoothly approximable, and ab and c' are independent there, the previous lemma applies and yields $a_c^{(1)} = b_c^{(1)}$ if and only if $a_{c,c'}^{(1)} = b_{c,c'}^{(1)}$; arguing similarly over c', our claim follows. In particular, E is a 0-definable equivalence relation.

Suppose toward a contradiction that is degenerate, i.e. $E = P^2$. Then for $c \in P$ fixed, the relation $E_c(a, b) : a_c^{(1)} = b_c^{(1)}$ has a class of maximal rank.

This violates the first clause of the previous lemma. As we are working over $acl(\emptyset)$, it follows that P/E is infinite. If c_1 is c/E, then $c_1 \in acl(c)$ and $rk(c/c_1) < rk(c)$. Therefore, by our initial assumption, $c_1 \in acl(\emptyset)$; that is, E has finite classes.

Let P have rank n and let $c_1, \ldots, c_{2n+1} \in P$ be independent. Let E_i be the equivalence relation $a_{c_i}^{(1)} = b_{c_i}^{(1)}$, and E' the intersection of the E_i. For any a, b in P, there is an i for which ab is independent from c_i and thus E' refines E, and has finite classes. Now $P/E' \leftrightarrow \prod_i P/E_i$, $\{c_1, \ldots, c_{2n+1}\}$-definably, and the quotients P/E_i are nonmultidimensional. Hence P is nonmultidimensional in $\mathcal{M}_{c_1, \ldots, c_{2n+1}}$. Therefore P is also nonmultidimensional over $\mathcal{M}(c_1)$, since any orthogonality over c_1 would be preserved (after conjugation) over $c_1, \ldots, c_{2n+1}$. As this case is the base of our induction, we are done. ∎

7.3 DEGENERATE GEOMETRIES

Lemma 7.3.1. *Let $\mathcal{M}$ be a structure and D 0-definable in $\mathcal{M}$. Then the following are equivalent:*

1. *D is stable and stably embedded in $\mathcal{M}$.*
2. *There is no unstable formula $\varphi(x, y)$ with $\varphi(x, y) \implies (x \in D)$.*
3. *There is no unstable formula $\varphi(x_1, \ldots, x_n, y)$ satisfying $\varphi(\mathbf{x}, y) \implies (x_i \in D)$, for all i.*

Proof. The equivalence of (2) and (3) is [Sh, II:2.13 (3, 4), p. 36]. We check the equivalence of (1) and (3).

Suppose first that (1) fails. If D is unstable then relativization to D produces a suitable φ. If D is not stably embedded and $\varphi(x, c)$ defines a subset of D which is not D-definable, one can find a countable set of parameters in D over which there are $2^{\aleph_0}$ φ^*-types (φ^* being φ with the variables interchanged). Indeed, for any finite set $A \subseteq D$ and any φ^*-type p over A realized by a conjugate of c, there are conjugates of c realizing contradictory φ-types over a larger finite subset of D; for this, we may suppose that p is satisfied by c, and take a 1-type over A in D which is split by $\varphi(x, c)$; then we have $\varphi(d_1, c)$ and $\neg\varphi(d_2, c)$ with d_1 conjugate to d_2 over A, and after identifying d_1 with d_2 we have realizations c, c' of contradictory φ-types by elements conjugate to c.

It follows that φ^* is unstable [Sh] [II:2.2 (1,2), pp. 30–31].

Now suppose (1) holds. Let A be a countable subset of D and $\mathcal{M}^*$ an elementary extension of $\mathcal{M}$. As D is stably embedded, any φ-type over D realized in $\mathcal{M}^*$ is definable with a parameter e in $D[\mathcal{M}^*]$, and since D is stable $tp(e/A)$ is definable. Thus the types over A are definable and (3) follows [Sh, II:2.2 (1, 8), pp. 30–31]. ∎

Lemma 7.3.2. *Let $\mathcal{M}$ be an $\aleph_0$-categorical structure which does not inter-pret a Lachlan pseudoplane. If $a, b \in \mathcal{M}$ with neither algebraic over the other, then there is a conjugate b' of b over a distinct from b for which $a \notin acl(b, b')$.*

Proof. Write down a theory asserting that $a_1, a_2, \ldots$ are distinct solutions to the conditions $tp(xb) = tp(xb') = tp(ab)$, with $b \neq b'$. Our claim is that this theory is consistent.

Suppose that this theory is inconsistent. Then for some n, b is definable from any n distinct conjugates $a_1, \ldots, a_n$ of a over b, by the conjunction of the formulas:

$$(*) \qquad\qquad tp(a_i, y) = tp(ab).$$

With n minimized (and at least 2) let $\mathbf{a} = \{a_1, \ldots, a_{n-1}\}$ be a set (unordered) of conjugates of a over b, chosen so that $b \notin acl(\mathbf{a})$. By assumption, none of the a_i is algebraic over b.

We claim that

1. $\mathbf{a} \notin acl(b)$.
2. $b \notin acl(\mathbf{a})$.
3. b is definable from any two distinct conjugates of $\mathbf{a}$ over b.
4. $\mathbf{a}$ is definable from any two distinct conjugates of b over $\mathbf{a}$.

Granted this, we have a Lachlan pseudoplane with points conjugate to $\mathbf{a}$, lines conjugate to b, and incidence relation given by $tp(\mathbf{a}b)$.

Now (1) is clear, (2) holds by the choice of $\mathbf{a}$ (and n), and for (3) observe that any two conjugates of $\mathbf{a}$ over b will involve at least n distinct conjugates of a over b. Finally, for (4), if b and b' have the same type over $\mathbf{a}$ and $\mathbf{a}, \mathbf{a}'$ are distinct and have the same type over bb', then b is definable from $\mathbf{aa}'$ in the manner of $(*)$ above, as is b', so $b = b'$. ∎

Definition 7.3.3. *A subset D of a structure $\mathcal{M}$ is* algebraically irreducible *if for $b \in D$ we have*

$$a \in acl(b) - acl(\emptyset) \text{ implies } b \in acl(a).$$

Lemma 7.3.4. *Let $\mathcal{M}$ be $\aleph_0$-categorical, let D be the locus of a 1-type over $\emptyset$ in $\mathcal{M}$, and suppose that D is algebraically irreducible and $\mathcal{M}$ does not interpret a pseudoplane. If there is a definable strongly minimal subset D_b of D with defining parameter b, then finitely many conjugates of D_b cover D.*

Proof. Let Q be the locus of b over $\emptyset$. Define an equivalence relation $E(b, b')$ on Q by the following condition: D_b and $D_{b'}$ differ by a finite set. By Lach-lan's normalization lemma [LaPP], for each $b \in Q$ there is a $D_{b/E}$-definable set agreeing with D_b up to a finite set. Thus we may factor out E and assume that distinct conjugates of D_b have finite intersection. Then the previous lemma applies to $a \in D_b - acl(b)$ and b, and as the conclusion fails, we find that for such pairs a, b we have $b \in acl(a)$. Now by the algebraic irreducibility of D it follows that $b \in acl(\emptyset)$. This yields our claim.   ∎

Lemma 7.3.5. *Let $\mathcal{M}^-$ be a reduct of the smoothly approximable structure $\mathcal{M}$. Let D be a rank 1 0-definable set in $\mathcal{M}^-$, and suppose that for any finite subset B of $\mathcal{M}^-$ and any a_1, a_2 in D: $acl(Ba_1a_2) = acl(Ba_1) \cup acl(Ba_2)$ where the algebraic closure is taken in D, and in the sense of $\mathcal{M}^-$. Then D is stable and is stably embedded in $\mathcal{M}^-$.*

Proof. Model theoretic notions are to be understood in $\mathcal{M}^-$ except where otherwise noted. The proof of (1) will proceed by induction on the rank r of D in $\mathcal{M}$. By Lemma 7.3.1 the class of stable and stably embedded 0-definable subsets of $\mathcal{M}^-$ is closed under finite unions. Thus we may suppose that D realizes a single type over $\emptyset$.

We show first that

Any infinite subset of D which is definable in $\mathcal{M}^-$ has rank r in $\mathcal{M}$.

Suppose, on the contrary, that D' is of lower rank in $\mathcal{M}$. Then by induction D' is stable and is stably embedded in $\mathcal{M}$ relative to a defining parameter for D'. From $\mathcal{M}$ D' inherits the following properties: it is $\aleph_0$-categorical, and does not interpret a pseudoplane. By Lachlan's theorem [LaPP] it is $\aleph_0$-stable and, in particular, contains a definable strongly minimal subset D'_b definable in $\mathcal{M}^-$. Then by the previous lemma finitely many conjugates of D'_b in $\mathcal{M}^-$ cover D and thus D is stable and stably embedded in $\mathcal{M}^-$.

From this it follows that for any sequence $a_1, a_2, \ldots$ in D which is algebraically independent in $\mathcal{M}^-$, there is a conjugate sequence which is independent in $\mathcal{M}$. Indeed, choosing the conjugates inductively, at stage n we have to realize the type of a_n over $a_1, \ldots, a_{n-1}$ in $\mathcal{M}^-$ (or more exactly a conjugate type) by an element independent from $a_1, \ldots, a_{n-1}$ in $\mathcal{M}$. The locus of this type is an infinite set defined in $\mathcal{M}^-$ and hence of full rank r in $\mathcal{M}$, so this is possible.

Now suppose we do not have D stable and stably embedded in $\mathcal{M}^-$, or equivalently that we have an unstable formula $\varphi(x, y)$ which implies $(x \in D)$. We then find a finite set B and types p, q over $acl(B)$ such that both $p(x), q(y), \varphi(x, y)$ and $p(x), q(y), \neg\varphi(x, y)$ have solutions with x, y independent over B. For this it suffices to take an indiscernible sequence (a_i, b_i) such that $\varphi(a_i, b_j)$ holds if and only if $i < j$, letting B be an initial segment over which the sequence is independent.

Now fix realizations b_{-1}, b_1 of q independent over B and set $B' = B \cup \{b_{-1}, b_1\}$. Let $D' = \{x \in D : \varphi(x, b_1) \& \neg\varphi(x, b_{-1})\}$. As $\mathcal{M}^-$ inherits the type amalgamation property from $\mathcal{M}$, by the corollary to Proposition 5.1.15 the set D' is infinite. Let $D'' \subseteq D'$ be the locus of a complete nonalgebraic type over B' in $\mathcal{M}^-$.

Now let $a_1, \ldots, a_n$ be elements of D'', pairwise algebraically independent over B'. We will show that there are 2^n φ-types over $a_1, \ldots, a_n$. By our basic assumption on D the set $A = \{a_1, \ldots, a_n\}$ is algebraically independent over B' and after conjugation we may suppose that these elements are independent in $\mathcal{M}$ over B'. For each i both $\varphi(a_i, y)\&q(y)$ and $\neg\varphi(a_i, y)\&q(y)$ are consistent, with rank equal to $rk(q)$, so by the corollary to type amalgamation the same applies to any combination of these properties as i varies. This produces the desired 2^n types.

Now let k be the size of $acl(B'a) \cap D$ in $\mathcal{M}^-$ for $a \in D''$. Then any set of n elements of D'' contains $[n/k]$ pairwise independent elements and hence allows $2^{[n/k]}$ φ-types. This is greater than the bound allowed by the corollary to Proposition 5.1.20. So we have a contradiction. ∎

Corollary 7.3.6. *With the hypotheses and notation of Lemma 7.3.5, if D carries no nontrivial 0-definable equivalence relation, then there is no induced structure on D beyond the equality relation.*

Proof. The additional hypothesis implies that $acl(a) = a$ for $a \in D$ and hence $acl(X) = X$ for $x \subseteq D$.

As we remarked in the previous proof, once we know that D is stable, we know that it is $\aleph_0$-stable and of Morley rank 1. By the Finite Equivalence Relation Theorem, the Morley degree is 1; that is, D is strongly minimal. As acl is trivial on D, the claim follows. ∎

7.4 REDUCTS WITH GROUPS

Lemma 7.4.1. *Let $\mathcal{M}^-$ be a reduct of a Lie coordinatizable structure $\mathcal{M}$, A a locally definable abelian group of bounded exponent n in $\mathcal{M}^-$. Then we have the following:*

1. *For any definable subset S of A, the subgroup generated by S is definable.*

2. *If A is 0-definable in $\mathcal{M}^-$ of exponent p, then the dual A^* and the pairing $A \times A^* \to F_p$ are interpretable in $\mathcal{M}^-$. If A has no nontrivial proper 0-definable subgroups in $\mathcal{M}^-$, then either A^* is trivial or the pairing is a perfect pairing.*

3. *If A is 0-definable and carries a 0-definable vector space structure over a finite field K, then A^* (the definable F_p-dual) allows a 0-definable K-bilinear pairing $\mu : A \times A^* \to K$ with $Tr \circ \mu(a, f) = f(a)$.*

Proof. These statements were proved in the Lie coordinatizable context as Lemma 6.1.8, Proposition 6.3.2, and Lemma 6.3.4.

The first statement is inherited from $\mathcal{M}$. The subgroup generated by S is definable in $\mathcal{M}$ if and only if it is generated in a finite number of steps, and this is equivalent to its definability in $\mathcal{M}^-$. Thus this first property passes to reducts.

For the second statement we have a definable dual $\hat{A}$ in $\mathcal{M}$, which, in particular, involves only finitely many sorts of $\mathcal{M}$, and we are interested in the subgroup A^* of $\mathcal{M}^-$-definable elements. Let A_n^* be the subset of $\mathcal{M}^-$-definable elements which are definable from at most n parameters. This generates a 0-definable subgroup of $\hat{A}$ and hence for large n is all of A^* in the sense of $\mathcal{M}^-$.

The proof of the third property is purely formal, given the second. ∎

Lemma 7.4.2. *Let $\mathcal{M}$ be a structure, and A a 0-definable abelian group in $\mathcal{M}^-$. Let H_i ($i = 1, \ldots, n$) be a finite set of subgroups of A, and let D be a finite union of cosets of the H_i, such that*

1. $[H_i : H_i \cap H_j]$ *is infinite for i, j distinct;*
2. *D contains a coset of each H_i and if D_i is the union of the cosets of H_i which are contained in D, there is no group $T > H_i$ commensurable with H_i for which D_i is the union of cosets of T.*

Then the groups H_i are $acl(\emptyset)$-definable in $(A; D)$.

Proof. This is an application of Beth's definability theorem applied to the set $\{H_1, \ldots, H_n\}$, which we claim is implicitly definable. Let n_i be the number of cosets of H_i contained in D_i and let $\mathcal{T}$ be the theory of (A, D) expanded by axioms φ for the H_i: they are subgroups with the stated properties, for which D_i is the union of exactly n_i cosets. Suppose we have two models of the form $(A, D, \bar{H})$ and $(A, D, \bar{H}')$ with the same (A, D). For each i, as some coset of H_i is covered by cosets of the H_j', by Neumann's lemma we have $[H_i : H_i \cap H_j'] < \infty$ for some j. Similarly for each j we can find a corresponding i; by the hypothesis on the H_i, these two correspondences are reciprocal, and after rearrangement this means that H_i is commensurable with H_i' for all i. Then for each i D_i is the same set in both models and is a union of cosets of both H_i and H_i', hence of $H_i + H_i'$; if this group extends H_i or H_i' properly, we contradict (2); but (2) can be included in φ since there is a bound on the possible index $[H_i + H_i' : H_i]$. Thus $H_i = H_i'$. ∎

Lemma 7.4.3. *Let $\mathcal{M}^-$ be a reduct of a Lie coordinatizable structure $\mathcal{M}$, and A a 0-definable abelian group in $\mathcal{M}^-$. Suppose that A has no definable subgroups in $\mathcal{M}^-$ of $\mathcal{M}$-rank strictly between 0 and $rk_{\mathcal{M}}(A)$. Then in $\mathcal{M}^-$, A has rank 1, and every infinite $\mathcal{M}^-$-definable subset of A has full rank in $\mathcal{M}$.*

Proof. The first statement follows from the second.

Suppose the second statement fails, and D is $\mathcal{M}^-$-definable in A with $0 < rk_\mathcal{M}(D) < rk_\mathcal{M}(A)$. Let $r = rk_\mathcal{M}(D)$ be minimal. By Lemma 6.2.5 in $\mathcal{M}$, D is contained in a finite union of cosets C_i of subgroups H_i of A definable in $\mathcal{M}$ with $rk\, H_i = r$, and a set of rank less than r. Let D be chosen to minimize the number n of distinct subgroups involved. Then the indices $[H_i : H_i \cap H_j]$ are infinite for i, j distinct.

We show that $n = 1$. By Lemma 6.2.5 $S_1 = Stab(D \cap C_1)$ has rank r, and evidently $S_1 \leq H_1$; but $rk\, H_1 = r$, so $[H_1 : S_1] < \infty$. Let a be a generic point of $Stab_\circ(D \cap C_1)$. Then $a \in H_1$ and $a \notin H_j$ for $H_j \neq H_1$, and furthermore $[a + C_j] \cap C_k = \emptyset$ for j, k distinct. Let $D' = D \cap (D + a)$; then $rk\, D' = r$ and up to a set of rank r D' is contained in the union of the $C_i \cap (C_j + a)$, which up to a set of rank less than r is the union of the cosets C_i for $H_i = H_1$. By the choice of D, the same applies to D and all $H_i = H$ coincide.

For $a \in A$ the set $D \cap (D + a)$ is $\mathcal{M}^-$-definable and hence is of rank r or finite. Thus $S_\circ = \{a \in A : rk(D \cap (D + a)) = r\}$ is definable in $\mathcal{M}^-$. Decompose D into loci of types D_i over $acl(\emptyset)$ in $\mathcal{M}$. Then $S_\circ = \bigcup_{ij} S_{ij}$ with $S_{ij} = \{a \in A : rk(D_i \cap (D_j + a)) = r\}$. By Lemma 6.2.5 each nonempty S_{ij} is contained in a coset C_{ij} of a subgroup T_{ij} of rank r, with $C_{ij} - S_{ij}$ of rank less than r. As D is contained in a finite union of cosets of H, also of rank r, H and the T_{ij} are commensurable.

Thus for some subgroup T of finite index in H, $S_\circ$ is a union of sets A_k contained in cosets of T and differing from these cosets by sets of rank less than r. Take $a_k \in A_k$ for each k, and let $Y_{kl} = (A_k - a_k) \cap (A_l - a_l)$. Then Y_{kl} is generically closed under addition and inverse, and applying Lemma 6.1.3, $A_k - A_l$ is a coset of a subgroup of T which differs from T by a set of smaller rank; so $A_k - A_l$ is a coset of T. From all of this it follows that $S_\circ - S_\circ$ is itself a finite union of cosets of T. As the set $S_\circ - S_\circ$ is definable in $\mathcal{M}^-$, the preceding lemma implies that some subgroup commensurable with T is also definable in $\mathcal{M}^-$. This contradicts our assumption on A. ∎

Lemma 7.4.4. *Let $\mathcal{M}^-$ be a reduct of a Lie coordinatizable structure $\mathcal{M}$, and A a rank 1 0-definable abelian group of prime exponent p in $\mathcal{M}^-$. Let A^* be the dual in $\mathcal{M}^-$ and let $\hat{A}$ be the dual in $\mathcal{M}$. Then:*

1. *In $\mathcal{M}^-$, A^* has rank at most 1.*
2. *If in $\mathcal{M}^-$ we have $acl(\emptyset) \cap A = (0)$, $acl(\emptyset) \cap A^* = (0)$, and $A^* \neq (0)$, then $A^* = \hat{A}$.*

Proof.

Ad 1. We apply the preceding lemma. Suppose A^* has a definable subgroup B in $\mathcal{M}^-$ with B and A^*/B infinite. Let $B^\perp$ be the annihilator of B in A. Then A^*/B acts faithfully on $B^\perp$, so $B^\perp$ is infinite. Similarly $(A/B^\perp, B)$ form a nondegenerate pair, so $A/B^\perp$ is infinite. This is a contradiction.

Ad 2. Let B be the annihilator in A of A^*. By hypothesis $B < A$ and hence $B = (0)$. Thus in $\mathcal{M}$ we have two perfect pairings (A, A^*) and $(A, \hat{A})$, and by the pseudofiniteness of $\mathcal{M}$ these dual groups coincide. ∎

Lemma 7.4.5. *Let $\mathcal{M}^-$ be a reduct of a Lie coordinatizable structure $\mathcal{M}$, A a rank 1 0-definable abelian group of prime exponent p in $\mathcal{M}^-$, and D an infinite 0-definable subset of A. Then for generic independent $a_1^*, \ldots, a_n^*$ in A^* there is $d \in D$ with (d, a_i^*) prescribed arbitrarily.*

Proof. By the last two lemmas every infinite $\mathcal{M}^-$-definable subset of A^* has full rank and thus the sequence $a_1^*, \ldots, a_n^*$ is conjugate in $\mathcal{M}$ to a generic independent sequence in A^*. Apply Lemma 6.4.1 in $\mathcal{M}$. ∎

Lemma 7.4.6. *Let $\mathcal{M}$ be a Lie coordinatizable structure, A a definable group abelian of rank r, and D a definable subset of A of rank r whose complement is also of rank r. Then there is a coset C of a definable subgroup of finite index in A, and an intersection D' of finitely many translates of D, such that*

$$rk(D') = r; \quad rk(D' \cap C) < r.$$

Proof. We may assume that A is settled over the empty set and that D is 0-definable. Let P be the locus of a 1-type over $acl(\emptyset)$. Then every definable subset of P is the intersection of P with a boolean combination of definable cosets of A of finite index, and of sets of rank less than $r = rk(P)$ (Lemma 6.6.2).

We may find a generic element $g \in A$ for which the rank of $P \setminus (D + g)$ is r: take $a \in A \setminus D$ generic, $b \in P$ generic with a, b independent, and $g = b - a$. There is a coset C of a definable subgroup of finite index in A, for which $C \cap P$ is contained in $P \setminus (D + g)$ up to a set of lower rank, or, in other words, $(D + g) \cap C \cap P$ has rank less than r. Furthermore, as A is settled over $acl(g)$, we may take C to be $acl(g)$-definable.

For each 1-type P over $acl(\emptyset)$ choose g_P and C_P as in the foregoing paragraph so that $\bigcap_P (D + g_P) \cap \bigcap_P C_P$ has rank less than r. Taking the g_P independent over the empty set, both intersections $\bigcap_P (D + g_P)$ and $\bigcap_P C_P$ will have rank r, and the latter is a coset of a definable subgroup of A of finite index. This proves the claim. ∎

Lemma 7.4.7. *Let $\mathcal{M}$ be a Lie coordinatizable structure, A a definable abelian group of rank r, and D a definable subset of A of rank r whose complement is also of rank r. Then there is an intersection D' of finitely many translates of D, which has rank r and is contained in a proper subgroup of finite index in A. In particular, the subgroup generated by D' will be a proper subgroup of finite index in A, which is definable in the structure (A, D).*

Proof. We apply the previous lemma to find a definable subgroup H of finite index in A, a coset C of H, and a finite intersection D' of finitely many translates of D, such that $D' \cap C$ has rank less than r. Take D' such an intersection, and suppose that the number of cosets of H which meet D' in a set of rank r is minimized, subject to the constraint that $rk\,D' = r$. We may suppose that $D = D'$: so if D meets $D + g$ in a set of rank r, then D and $D + g$ meet the same cosets of H in a set of rank r.

Let $X \subseteq A/H$ be the set of cosets which meet D in a set of rank r. We may suppose $H \in X$.

We claim that X is a subgroup of A/H. We may take D and H to be 0-definable. Take $C \in X$ and choose a representative g for C as follows. Fix a 1-type over $acl(\emptyset)$ whose locus P is contained in $D \cap H$, and let Q be the locus of a 1-type over $acl(\emptyset)$ which is contained in $D \cap C$. Take $(a, b) \in P \times Q$ generic; then $g = b - a$ is generic, and $g + H = C$. Furthermore, $(g + D) \cap D \cap Q$ contains $(g + P) \cap Q$ (in particular, a) and hence has full rank. Thus $g + D$ also meets all the cosets in X in sets of rank r, in other words $X - g = X$. Thus X is a group.

Let $X = B/H$ with $H \le B \le A$. As $C \notin X$, we have $B < A$. In addition, by our construction $D \backslash B$ has rank less than r. Let $S = D \backslash B$. As $rk\,S < r$, for any $r + 1$ independent generic elements $h_1, \ldots, h_{r+1}$ in A we will have $\bigcap_i (S + h_i) = \emptyset$; if c lies in the intersection and is independent from h_i, then $rk(h_i/c) = r$, and $c - h_i \in S$, a contradiction.

Thus if we replace D by the intersection D' of its translates by $r + 1$ independent generic elements of B, then we will retain $rk\,D' = r$, while now $D' \subseteq B$. ∎

Proposition 7.4.8. *Let $\mathcal{M}^-$ be a reduct of a Lie coordinatizable structure $\mathcal{M}$, A a rank 1 0-definable group in $\mathcal{M}^-$. If $A^* = (0)$ in $\mathcal{M}^-$, then A is strongly minimal and stably embedded in $\mathcal{M}^-$.*

Proof. Supposing the contrary, there is a subset D of A which is definable in $\mathcal{M}^-$ (from parameters in $\mathcal{M}^-$), is infinite, and has infinite complement. By Lemma 7.4.3, both D and its complement have full rank in A. By Lemma 7.4.7 there is a proper subgroup of finite index in A which is definable in $\mathcal{M}^-$; so A^* is nontrivial in $\mathcal{M}^-$. ∎

Proposition 7.4.9. *Let $\mathcal{M}^-$ be a reduct of a Lie coordinatizable structure $\mathcal{M}$, A a rank 1 0-definable group in $\mathcal{M}^-$. Suppose $acl(\emptyset) \cap A = (0)$, and $acl(\emptyset) \cap A^* = (0)$. Then there is a finite field F and an $acl(\emptyset)$-definable F-space structure on A for which algebraic closure on A and F-linear span coincide.*

Proof. We let F be the ring of $acl(\emptyset)$-definable group endomorphisms of A, which is a division ring and is finite by $\aleph_0$-categoricity; thus it is a finite field.

We show by induction on n that any n F-linearly independent elements of A are independent. Assuming the claim for n, suppose that $a \in acl(a_1, \ldots, a_n)$ with $a_1, \ldots, a_n$ independent. We claim that a is a linear combination of the a_i. Taking a conjugate of $a_1, \ldots, a_n$ in $\mathcal{M}^-$ we may suppose that the elements $a_1, \ldots, a_n$ are independent of maximal rank in $\mathcal{M}$.

Consider the locus D of $a_1, \ldots, a_n, a$ over $acl(\emptyset)$ in $\mathcal{M}$, and let S be $Stab(D)$. By Lemma 6.2.5 $rk\,S = rk\,D = n \cdot rk_\mathcal{M}(A)$, and D is contained in a coset of S. Let T be the projection of S onto the first n coordinates. Then the projection of D is contained in a coset of T and thus $rk\,T = rk\,S$. Therefore the kernel is finite, and T has finite index in A^n. We claim:

$$(*) \qquad \text{Some subgroup } S' \text{ of } A^n \text{ commensurable with } S \text{ is } acl(\emptyset)\text{-definable in } \mathcal{M}^-.$$

For any $\mathcal{M}^-$-definable subset X of A^n one sees easily by induction on n that $rk_\mathcal{M} X = rk\,X \cdot rk_\mathcal{M} A$. Accordingly $Stab_\circ(D)$ in the sense of $\mathcal{M}$ is definable in $\mathcal{M}^-$. One then continues as in the final paragraph of the proof of Lemma 7.4.3. Thus $(*)$ holds.

In $\mathcal{M}$, $S' \cap S$ is also $acl(\emptyset)$-definable and induces an equivalence relation on D with finitely many classes. As D is complete over $acl(\emptyset)$ in $\mathcal{M}$, it is contained in a single coset of $S \cap S'$ and thus $S \leq S'$ with $[S' : S] < \infty$. The kernel of the projection of S' to the first n coordinates is also finite, hence trivial by our hypotheses, and the image is of finite index in A^n, hence the projection is surjective. It follows that S' represents a linear function $s(\mathbf{x}) = \sum_i \alpha_i x_i$ with coefficients in F. As D lies in a coset of S, it lies in a coset of S', and the function $y - s(\mathbf{x})$ is constant on D, hence in $acl(\emptyset)$ in $\mathcal{M}^-$, hence 0. Thus $h(a) = \sum_i \alpha_i a_i$. ∎

Lemma 7.4.10. *Let $\mathcal{M}$ be $\aleph_0$-categorical and modular of finite rank, $\mathcal{M}^-$ a reduct of $\mathcal{M}$ with $acl_\mathcal{M}(\emptyset) = acl_{\mathcal{M}^-}(\emptyset)$. If X, Y are sets which are independent in $\mathcal{M}$, then they are independent in $\mathcal{M}^-$.*

Proof. If X, Y are dependent in $\mathcal{M}^-$ then in $\mathcal{M}^-$ by inherited modularity there is $a \in acl(X) \cap acl(Y) - acl(\emptyset)$ and by our hypothesis this holds also in $\mathcal{M}$. ∎

Proposition 7.4.11. *Let $\mathcal{M}^-$ be a reduct of a Lie coordinatizable structure $\mathcal{M}$, A a rank 1 0-definable group in $\mathcal{M}^-$, and suppose that $acl_\mathcal{M}(\emptyset) \cap (\mathcal{M}^-)^{eq} = dcl_{\mathcal{M}^-}(\emptyset)$. If A is settled over $\emptyset$ in $\mathcal{M}$, then it is settled over $\emptyset$ in $\mathcal{M}^-$ and thus every definable subset in $\mathcal{M}^-$ is a boolean combination of 0-definable subsets, a finite set, and cosets of definable subgroups.*

Proof. We must show in $\mathcal{M}^-$ that for a independent from an algebraically closed set c,

$$tp(a) \cup gtp(a/c \cap A^*) \Longrightarrow^* tp(a/c)$$

(all types are computed in $\mathcal{M}^-$). We will show in fact that for *any* c there is a linearly independent k-tuple $\mathbf{b} \in A^*$ for some k for which

$(*)$
$$tp_{\mathcal{M}^-}(a) \cup gtp(a/\mathbf{b}) \Longrightarrow tp_{\mathcal{M}^-}(a/c)$$
$$\text{for any } a \in A \text{ not algebraic over } b, c.$$

After absorbing those parameters in $\mathbf{b}$ which are algebraic over c into c, the rest are independent over c and are conjugate in $\mathcal{M}^-$ to parameters independent in $\mathcal{M}$ over c. For a independent from c in $\mathcal{M}^-$ with $gtp(a/c \cap A^*)$ as specified, we can conjugate a over c to an independent element in $\mathcal{M}$, then by type amalgamation complete a, c to $a, \mathbf{b}, c$ with the same 2-types $tp(a\mathbf{b})$ and $tp(\mathbf{b}c)$ as in the original triple $a, \mathbf{b}, c$ (that is, the version in which $\mathbf{b}$ is independent from c). This then determines $tp(a/c)$. Note that in the course of the argument a portion of $acl(c \cap A^*)$ was absorbed into c.

We now begin the lengthy verification of $(*)$.

Let C be the locus of the type of c over $\emptyset$ in $\mathcal{M}^-$ and let k be the maximum dimension of $acl_{\mathcal{M}}(c) \cap A^*$ for $c \in C$. Let B_k be the set of linearly independent k-tuples in A^*. We introduce the notation "$cl(c)$" for the set $\{a \in A : rk_{\mathcal{M}}(a/c) < rk_{\mathcal{M}}(A)\}$.

We consider the following two relations E^-, E on pairs from $B_k \times C$. $E^-((b, c), (b', c'))$ holds if and only if (b, c) is independent from (b', c') in $\mathcal{M}^-$ and for $a \in A - acl(b, b', c, c')$, we have $gtp(a/b) = gtp(a/b')$ implies $tp(a/c) = tp(a/c')$; $E((b, c), (b', c'))$ holds if and only if (b, c) is independent from (b', c') in $\mathcal{M}$ and for $a \in A - cl(b, b', c, c')$, we have $gtp(a/b) = gtp(a/b')$ implies $tp(a/c) = tp(a/c')$.

Then easily E holds if and only if E^- holds and the pairs involved are independent in $\mathcal{M}$. Now we show that E is a generic equivalence relation in the sense of §5.1. So take an independent triple $x = (b, c)$; $x' = (b', c')$; $x'' = (b'', c'')$ in $\mathcal{M}$, with $E(x, x')$ and $E(x', x'')$ holding. We must show $E(x, x'')$.

Take $a \in A - cl(b, b'', c, c'')$ with $gtp(a/b) = gtp(a/b'')$. We claim $tp(a/c) = tp(a/c'')$. Let $q = tp(a)$, $r = gtp(a/b) = gtp(a/b'')$. By Lemma 6.4.1 $q(x) \cup r(x/b')$ is consistent, of rank $rk\,q$. By the corollary to type amalgamation (Proposition 5.1.15), the same holds for the type

$$q(x) \cup r(x/b') \cup \underset{\mathcal{M}}{tp}(a/bb''cc'').$$

Take $a' \in A - cl_{\mathcal{M}}(bb'b''cc'c'')$ realizing this type. From $E(x, x')$ and $E(x', x'')$ we find in $\mathcal{M}^-$ that $tp(a'c) = tp(a'c') = tp(a'c'')$, and thus $tp(ac) = tp(ac'')$.

Now we claim that E^- is also a generic equivalence relation. Let x, x', x'' be independent in $\mathcal{M}^-$ with $E^-(x, x')$ and $E^-(x', x'')$. We can conjugate x, x', x'' in $\mathcal{M}^-$ to an independent triple in $\mathcal{M}$ and reduce to the case of E.

Accordingly by Lemma 5.1.12 there is a 0-definable equivalence relation E' in $\mathcal{M}^-$ that agrees with E^- on independent pairs in $B_k \times C$. Then E' also agrees with E on $\mathcal{M}$-independent elements of $B_k \times C$. The domain of the relation E' is $D =:$

$$\{x \in B_k \times C \ : \ \text{There is } x' \in B_k \times C \text{ independent from } x$$
$$\text{such that } E^-(x, x')\}.$$

Note that in this definition we may take independence in the sense either of $\mathcal{M}$ or of $\mathcal{M}^-$ since these notions agree up to conjugation in $\mathcal{M}^-$.

We consider also the following set, which will turn out to coincide with D:

$$D_1 = \quad \{(b, c) \in B_k \times C : \text{For } a \in A - cl(b, c),$$
$$tp_{\mathcal{M}}(a) \cup gtp(a/b) \text{ determines } tp_{\mathcal{M}^-}(a/c)\}.$$

Note that if b includes a basis for $acl(c) \cap A^*$, then as A is settled in $\mathcal{M}$, $(b, c) \in D_1$. Thus D_1 projects onto C. Furthermore, E has finitely many classes on D_1 since for $x \in D_1$, the class of x/E' is determined by information in $tp_{\mathcal{M}}(x')$. (This is clear first for independent pairs x, x' using the definition of E^- and then for general pairs.)

We will show shortly that $D = D_1$. First we check that D projects onto C. Take $c \in C$, and b linearly independent containing a basis for $acl_{\mathcal{M}}(c) \cap A^*$. Take a conjugate (b', c') in $\mathcal{M}$ independent from (b, c) in $\mathcal{M}$. Then easily $E((b, c), (b', c'))$ and thus $(b, c) \in D$. By the same argument $D_1 \subseteq D$.

We will now show $D \subseteq D_1$. Let $x \in D$, and x' independent from x in $\mathcal{M}$, with $E(x, x')$. With $x = (b, c)$ we must show that $tp_{\mathcal{M}}(a) \cup gtp(a/b)$ determines $tp_{\mathcal{M}^-}(a/c)$ for $a \in A - cl(x)$. Let $a, a' \in A - cl(x)$ satisfy $tp_{\mathcal{M}}(a) = tp_{\mathcal{M}}(a') = q$ and $gtp(a/b) = gtp(a'/b) = r(x/b)$. By type amalgamation we may choose a, a' so that the triple $a; a'; bb'cc'$ is independent in $\mathcal{M}$ and a and a' satisfy the same type over $b'c'$. This then yields $tp_{\mathcal{M}^-}(a/c) = tp_{\mathcal{M}^-}(a/c') = tp_{\mathcal{M}^-}(a'/c') = tp_{\mathcal{M}^-}(a'/c)$. Thus $(b, c) \in D_1$.

Finally, we prove $(*)$. The relation E' has finitely many classes on $D_1 = D$. As $acl(\emptyset) = dcl(\emptyset)$ any such class $D_\circ$ is 0-definable in $\mathcal{M}^-$. Let $(b, c) \in D_\circ$ and suppose that our claim fails for (b, c). Fix $a, a' \in A - acl(b, c)$, with equal types in $\mathcal{M}^-$ and with $gtp(a/b) = gtp(a'/b)$ but with $tp_{\mathcal{M}^-}(a/c) \neq tp_{\mathcal{M}^-}(a'/c)$. Let σ be an automorphism carrying a' to a. Then $gtp(a/b) = gtp(a/\sigma b)$ but $tp_{\mathcal{M}^-}(a/c) \neq tp_{\mathcal{M}^-}(a/\sigma c)$.

Take a pair (b', c') conjugate to (b, c) over a in $\mathcal{M}^-$ which is independent from $b, c, \sigma b, \sigma c$. Then

$$gtp(a/\sigma b) = gtp(a/b) = gtp(a/b')$$

and $tp_{\mathcal{M}^-}(a/c') \neq tp_{\mathcal{M}^-}(a/\sigma c)$. As $(\sigma b, \sigma c)$ and $(b'c')$ are independent, this shows they are inequivalent with respect to E. However, these pairs are conjugate in $\mathcal{M}^-$, a contradiction. ∎

Corollary 7.4.12. *Let $\mathcal{M}^-$ be a reduct of a Lie coordinatizable structure $\mathcal{M}$, A a rank 1 0-definable group in $\mathcal{M}^-$. If A is settled over $\emptyset$ in $\mathcal{M}$ then it is settled in $\mathcal{M}^-$ over a finite set of $\mathcal{M}$-algebraic constants.*

Proof. By the preceding result A becomes settled over $acl_{\mathcal{M}}(\emptyset)$ and hence over the collection of definable subsets of A which belong to $acl_{\mathcal{M}}(\emptyset)$; there are finitely many such. ∎

7.5 REDUCTS

In the present section we show that reducts of Lie coordinatized structures are weakly Lie coordinatized; we may lose the orientation. We must deal mainly with the primitive case (meaning there is no nontrivial 0-definable equivalence relation).

Lemma 7.5.1. *Let $\mathcal{M}$ be a structure realizing finitely many 3-types, and $a \in \mathcal{M}$. Let $acl(a)$ be computed in $\mathcal{M}^{eq}$. Then the lattice of algebraically closed subsets of $acl(a)$ is finite.*

Proof. Let $\mathcal{E}_a$ be the collection of a-definable equivalence relations on $\mathcal{M}$ which have finitely many classes, $C_a = \bigcup\{\mathcal{M}/E : E \in \mathcal{E}_a\}$, and $\hat{C}_a$ the collection of subsets of $\mathcal{M}$ which are unions of subsets of C_a. Viewing $\hat{C}_a$ as a subset of $\mathcal{M}^{eq}$, we have $\hat{C}_a \subseteq acl(a)$, and it suffices to show that for $\alpha \in acl(a)$ we have

$$(*) \qquad\qquad \alpha \in acl(acl(\alpha) \cap \hat{C}_a).$$

Let $\alpha \in acl(a)$ and let $\varphi(x, a)$ be a formula which defines a finite set A containing α. Let $S = \{b \in \mathcal{M} : \varphi(\alpha, b)\}$, which we view as an element of $\mathcal{M}^{eq}$, and let $A_S = \{\beta : \forall x \in S \; \varphi(\beta, x)\}$. Then easily $S \in dcl(\alpha) \cap \hat{C}_a$, and as $\alpha \in A_S \subseteq A$, we have $\alpha \in acl(S)$. This proves $(*)$ (and a little more). ∎

Remark 7.5.2. When $\mathcal{M}$ is $\aleph_0$-categorical, the foregoing lemma applies to any element a of $\mathcal{M}^{eq}$. (For another approach, see the note at the end of this section.)

Proposition 7.5.3. *Let $\mathcal{M}$ be a weakly Lie coordinatized structure, $\mathcal{M}^-$ a reduct of $\mathcal{M}$, and D a primitive, rank 1, definable subset of $\mathcal{M}^-$. Then D is a Lie geometry forming part of a Lie geometry stably embedded in $\mathcal{M}^-$; this geometry may be unoriented, and may be affine.*

Proof. As D has rank 1, acl gives a combinatorial geometry on D; the same holds over any finite set.

Suppose first that acl_B gives a degenerate geometry over any finite B, or in other words, that $acl(A, B) = \bigcup_{a \in A} acl(a, B)$ in D. In this case, by Lemma 7.3.5, D is a trivial structure, and is stably embedded.

Now we deal with the nondegenerate case. Let $\{D_i\}$ be a set of representatives for the primitive rank 1 $acl(\emptyset)$-definable sets in D^{eq}, up to 0-definable bijections, with $D_1 = D$, and let $D^\infty = \bigcup_i D_i$. We claim that D^∞, with acl, is a projective space (of infinite dimension) over a field; the field will be finite by the previous lemma, applied as indicated in the subsequent remark.

We show first that some line has more than two points. Take c_1, c_2, c_3 in D and B a finite set such that $c_3 \in acl(c_1 c_2 B) - [acl(c_1 B) \cup acl(c_2 B)]$. By modularity there is $e \in acl(c_1 c_2) \cap acl(c_3 B)$ such that $c_1 c_2$ and $c_3 B$ are independent over e. Then $rk(e) = 1$ and we may take $e \in D^\infty$. As $e \in acl(c_1, c_2) - [acl(c_1) \cup acl(c_2)]$, this suffices.

Now we show that coplanar lines meet. Take a_1, a_2, a_3, a_4 in D^∞ pairwise algebraically independent with $rk(a_1 a_2 a_3 a_4) = 3$. Take e in $acl(a_1 a_2) \cap acl(a_3 a_4)$ such that $a_1 a_2$ and $a_3 a_4$ are independent over e. Then again $rk\, e = 1$ and e may be taken in D^∞.

Thus D^∞ is an infinite-dimensional projective geometry with finite lines, and there is a vector space model, that is a map $\pi : V - (0) \to D^\infty$ in which linear dependence in V corresponds to algebraic independence in D^∞. We do not claim that this vector space is interpreted globally in the model.

Let $V_i = \pi^{-1}[D_i]$, thought of as a new sort for each i. We enrich $\mathcal{M}^-$ by the V_i with the relevant structure, taking π_i to be the restriction of π to V_i, and restricting $+$ and scalar multiplication to a family of relations on the new sorts. The expanded structure will be called $\mathcal{M}^{-*}$; it can be thought of also as a reduct $\mathcal{M}^{*-}$ of an expansion of the original structure $\mathcal{M}$ by the new sorts and relations. Here $\mathcal{M}^*$ is a finite cover of $\mathcal{M}$ by sets of order $q - 1$; any automorphism of $\mathcal{M}$ over $acl(\emptyset)$ extends to an automorphism of $\mathcal{M}^*$. Thus $\mathcal{M}^*$ is weakly Lie coordinatizable.

By Lemma 7.4.1 V_1 lies in a 0-definable rank 1 group A in $\mathcal{M}^{-*}$. We may suppose that A has no 0-definable finite subgroups. Our claim is that

$(*)$ A is part of a stably embedded Lie geometry in $\mathcal{M}^{-*}$.

Assuming $(*)$, D forms part of an embedded Lie geometry J in $\mathcal{M}^-$; the induced structure may be computed in $\mathcal{M}^{-*}$. Furthermore, the geometry in $\mathcal{M}^{-*}$ is algebraic over J; A is algebraic over D and if for example A^* is nontrivial then it is algebraic over its projectivization, which is in J. Thus J is stably embedded in $\mathcal{M}^{-*}$ and *a fortiori* in $\mathcal{M}^-$: for $e \in \mathcal{M}^{-*}$, $tp(e/A)$ is definable by parameters $a \in A$, whose type over J is algebraic and hence definable. Thus it suffices to prove $(*)$.

Suppose first that A has no 0-definable proper subgroup of finite index. If $A^* = (0)$ in $\mathcal{M}^{-*}$ then Proposition 7.4.8 applies. Otherwise, A^* is the full definable linear dual to A, also in $\mathcal{M}^*$, by Lemma 7.4.4. A and A^* are settled over some parameter c in $\mathcal{M}^*$, hence in $\mathcal{M}^{-*}$ settled over some parameter algebraic in c by the corollary to Proposition 7.4.11. After enlarging c further we may suppose that $acl(c) \cap (A, A^*)$ also carries a nondegenerate pairing and lies in $dcl(c)$. By Lemma 7.4.5 and Proposition 7.4.9, Proposition 7.1.7 applies.

Now suppose that A does have a proper 0-definable subgroup of finite index; let B be the least such. Then by the preceding paragraph B is part of a stably embedded Lie geometry (B, B^*, Q), some components of which may be empty. A is generated by a complete type whose image in A/B must be a single point. Thus the dimension of A/B is 1. Then (A, B, B^*, Q) may be viewed as an affine geometry, by Lemma 2.3.17(1), with $C = \emptyset$. ∎

Below we give another treatment of the degenerate case on somewhat different lines.

Proposition 7.5.4. *Let $\mathcal{M}^-$ be a reduct of a Lie coordinatized structure. Then $\mathcal{M}^-$ is weakly Lie coordinatized.*

Proof. $\mathcal{M}^-$ is $\aleph_0$-categorical, has finite rank, is modular, and enjoys the following additional property:

If $a, b \in \mathcal{M}^-$, $a \notin acl(b)$, then there is $a' \in acl(a)$ of rank 1 over b.

This is contained in Lemma 5.6.6. Thus for any $a \in \mathcal{M}$ we can find a chain of "coordinates" $a_1, \ldots, a_n$ of finite length with a_i belonging to a rank 1 primitive $acl(a_{i-1})$-definable set D_i and $a_n = a$. By Proposition 7.5.3 D_i is part of a stably embedded Lie geometry and after interposing the algebraic parameters needed to define the D_i we obtain a weak Lie coordinatization. ∎

We now return to the degenerate case, indicating a treatment based on weaker hypotheses. We refer here to the preprint [HrS1], which introduced the S_1 *rank* on formulas as the least rank subject to:

($*$) $S_1(\varphi) > n$ iff there are $(b_i)_{i \in \mathbb{N}}$ indiscernible over a set of definition for φ, and a formula $\varphi'(x, y)$, such that
1. $S_1(\varphi \& \varphi'(x, b_i)) \geq n$ for each i;
2. For some k: $S_1(\varphi'(x, b_1) \& \ldots \& \varphi'(x, b_k)) < n$.

The independence theorem can be proved for theories of finite S_1 rank by an argument isomorphic to the one which will be given at the end of §8.4.

Lemma 7.5.5. *Let M be an $\aleph_0$-categorical structure of finite rank with amalgamation of types, not interpreting the generic bipartite graph, and let $\mathcal{M}^-$ be a reduct of M. Let D be a primitive rank one definable subset in $\mathcal{M}^-$ whose geometry is orthogonal to every primitive rank 1 set whose geometry is nondegenerate; in particular D is degenerate over any finite set. Then D is stably embedded and trivial.*

Proof. Any rank 1 subset of $\mathcal{M}^-$ will inherit from $\mathcal{M}$ the property of finite S_1-rank, and hence satisfy the type amalgamation property by [HrS1].

To see that D is stably embedded and trivial we will show that for any finite B, D remains primitive over $D - acl(B)$. For this we may use induction on $rk\,B$, and thus by analyzing B we may suppose that $B = \{b\}$ has rank 1. Let D' be the locus of b over $\emptyset$, a rank 1 set. Let E be a b-definable equivalence relation on $D - acl(b)$. As D is degenerate this will not have finite classes, so it will have finitely many infinite classes. Suppose $a_1, a_2 \in D - acl(b)$ are distinct and equivalent, while a_1', a_2' are inequivalent. Then b, a_1, a_2 are pairwise independent, as are b, a_1', a_2', and hence independent. If a_1, a_2, and a_1' all have the same type over $acl(b)$ then amalgamating types over $acl(b)$ we can find $a_1^*, a_2^*, a_1'^*$ realizing this type with $tp(ba_1^*a_2^*) = tp(ba_1a_2) = tp(ba_1^*a_1'^*)$, and $tp(ba_2^*a_1'^*) = tp(ba_1'a_2')$. Then a_1^* is E-equivalent to a_2^* and $a_1'^*$ but they are not E-equivalent to each other, a contradiction.

Thus $D - acl(b)$ splits into at least two types over $acl(b)$. In particular D carries a nontrivial equivalence relation definable from the set $acl(b)$ (or a part of it meeting finitely many sorts), viewed as a single element of $\mathcal{M}^{-eq}$. This being the case, we may replace D' by a primitive quotient, and the argument of the previous paragraph yields a 0-definable relation $R(x, y)$ on $D' \times D$ such that $R(b, y)$ splits $D - acl(b)$ for $b \in D'$. We view (D', D) as a bipartite graph with edge relation R. By our hypothesis D' also carries a degenerate geometry.

As $R(b, a)$ and $\neg R(b, a)$ both occur with $a \notin acl(b)$, by amalgamation of types any two finite subsets of D can be separated by an element of D', and similarly for D' over D. Thus this is the generic bipartite graph, a contradiction. ∎

We now return to Theorem 7 of §1.2.

Theorem 7.5.6 (Theorem 7: Model Theoretic Analysis)

The weakly Lie coordinatizable structures $\mathcal{M}$ are characterized by the following nine model theoretic properties.

LC1. $\aleph_0$-*categoricity.*

LC2. *Pseudofiniteness.*

LC3. *Finite rank.*

LC4. *Independent type amalgamation.*

LC5. *Modularity of $\mathcal{M}^{eq}$.*

LC6. *The finite basis property for definability in groups.*

LC7. *Lemma 6.4.1: we call this "general position of large 0-definable sets."*

LC8. *$\mathcal{M}$ does not interpret the generic bipartite graph.*

LC9. *For every vector space V interpreted in $\mathcal{M}$, the definable dual V^* (the set of all definable linear maps on V) is interpreted in $\mathcal{M}$.*

Proof. One has to check in the first place that these properties hold in weakly Lie coordinatizable structures. These statements have been proved in various earlier sections. Note however that the properties (LC6) and (LC7) were treated in the Lie coordinatizable context. As noted at the outset in §6.3, any group interpreted in a weakly Lie coordinatizable structure is also interpreted in a Lie coordinatizable structure, so these properties also apply in the weakly Lie coordinatizable context.

For the converse, observe that we have listed here most of the properties used in the analysis of reducts of Lie coordinatized structures, with the noteworthy exception of aspects of the theory of envelopes. We need to see that the proof of Proposition 7.5.4 can be carried out in this context.

This proposition depends on Proposition 7.5.3 and Lemma 5.6.6; the latter holds in our context, so we need only concern ourselves with Proposition 7.5.3. The use of Lemma 7.3.5 in the proof of that proposition does not fit into the present context, and it must be replaced by Lemma 7.5.5, using hypothesis (LC9) to see that the orthogonality condition in Lemma 7.5.5 will hold for any geometry D which is degenerate over every finite set. In a wider context, it is possible for a set to act as a generic set of linear maps on a vector space, giving a bipartite structure reminiscent of both the generic bipartite graph and the polar geometry; in this case, one would have a degenerate geometry nonorthogonal to a linear geometry, and, in fact, embedded in the definable dual (which, however, would not itself be interpretable.) Condition (LC9) and nonorthogonality imply that over some parameter set, $acl(D)$ contains an infinite definable group; we leave the details of this (involving the definition of orthogonality as well as the nature of the definable sets in a nondegenerate geometry) to the reader.

So it remains to verify that the rest of the proof of Proposition 7.5.3, which makes use of a large body of machinery, is available in the context of properties (LC1–LC9). The ingredients of Proposition 7.5.3, apart from (LC1, LC3, LC5), are the following: a particular finite covering of $\mathcal{M}^-$; Lemmas 2.3.17 and 6.6.2; Propositions 6.6.1 and 7.1.7; the contents of §7.4.

Properties (LC1-LC5) are inherited directly by the cover. Properties (LC6, LC7) can be deduced by showing that the groups interpreted in the cover are also interpretable in $\mathcal{M}^-$. This is because each sort (V_i) in the cover is interpretable in (part of) the underlying projective geometry: fix two linearly independent vectors v_1, v_2 and associate with any linearly independent v the pair $\langle v - v_1 \rangle, \langle v - v_2 \rangle$.

Lemma 2.3.17 simply holds, and Lemma 6.6.2 holds for the case needed by (LC7). Proposition 6.6.1 is assumption (LC6) and Proposition 7.1.7 was proved under our assumptions. So it suffices to reexamine §7.4. Lemma 7.4.1

may be replaced by Lemma 6.1.8 in the present context. The remaining lemmas, down to Lemma 7.4.7, are available in our context; note that Lemma 7.4.3 depends on lemmas in §§6.1-6.2 which were proved under sufficiently general hypotheses. Then the proofs of Propositions 7.4.9 and 7.4.11 can be repeated. We do not need Proposition 7.4.11 since we assume (LC6).   ∎

Note. The following alternative route to the finiteness statement needed for the proof of Proposition 7.5.3 (Lemma 7.5.1 and the subsequent remark) has its own interest:

Lemma 7.5.7. *If $\mathcal{M}$ is saturated and $a \in \mathcal{M}$, then every algebraically closed subset of $acl(a)$ is of the form $acl(a) \cap acl(a')$ for some conjugate a' of a in $\mathcal{M}$.*

Proof. Let $A \subseteq acl(a)$ be algebraically closed. We need to check the consistency of the following theory, involving a new constant c and constants for the elements of A:

$$tp(c/A) = tp(a/A); \; b \notin acl(c) \text{ (for } b \in acl(a)\backslash A).$$

For this it suffices to check for each finite a-definable subset B of $acl(a)$ that there is an automorphism α of $\mathcal{M}$ fixing A such that

$(*)$ $(B\backslash A) \cap (B\backslash A)^\alpha = \emptyset$

Let $G = Aut(\mathcal{M})_A$, the pointwise stabilizer of A in $Aut(\mathcal{M})$. For $b_1, b_2 \in B\backslash A$, let $G(b_1, b_2) = \{\alpha \in G : b_1^\alpha = b_2\}$. This is a coset of G_{b_1}, and if G is covered by $G(b_1, b_2)$ as b_1, b_2 vary over $B\backslash A$, then by Neumann's Lemma one of the subgroups G_b ($b \in B\backslash A$) has finite index in G; but this means $b \in acl(A) = A$, a contradiction. Thus condition $(*)$ can be met.   ∎

8

Effectivity

8.1 THE HOMOGENEOUS CASE

If $\mathcal{M}$ is a finite relational language let $\mathcal{M}^\infty$, or more properly $\mathcal{M}^{\infty,\mathrm{eq}}$, be the language augmented by the quantifier $\exists^\infty$ "there exist infinitely many," and expanded so as to apply to imaginary elements.

We consider the following effectivity problems.

Problems

    (A) Given a finite relational language $\mathcal{M}$ and a sentence φ in the language $\mathcal{M}^\infty$, is there a stable homogeneous model (of type $\mathcal{M}$) of φ?

    (B) Given a finite relational language and a finite set of forbidden isomorphism types $\mathcal{C}$, consisting of isomorphism types of finite $\mathcal{M}$-structures, is the corresponding class $\mathcal{A}(\neg\mathcal{C})$ an amalgamation class with stable generic structure? Here $\mathcal{A}(\neg\mathcal{C})$ denotes the class of finite structures omitting the structures of type $\mathcal{C}$.

A restricted version of Problem A was considered by Knight and Lachlan in [KL], and treated in the binary case. As there is an a priori bound on the rank in this case, the question is one of the consistency of a theory in the extended language, hence a negative answer will have a finite verification.

The idea of [KL] is to reduce the positive case to Problem B. If $\mathcal{M}$ is a stable homogeneous model satisfying φ and $\mathcal{C}$ is the class of minimal isomorphism types of structures omitted by $\mathcal{M}$, then $\mathcal{C}$ is finite, as a consequence of the quasifinite axiomatizability. Thus $\mathcal{C}$ is a finite object witnessing the existence of $\mathcal{M}$, and the problem is to recognize $\mathcal{C}$.

If N bounds the sizes of the constraints in $\mathcal{C}$, then the quantifier $\exists^\infty$ is equivalent to $\exists^{N^*}$, where N^* is so large that every $\mathcal{M}$-structure of size N^* contains an indiscernible sequence of size N. This reduces the problem to the first order case. As $\mathcal{C}$ determines a "quantifier elimination" procedure—where the quotation marks reflect a bad conscience in cases where there is, in fact, no associated homogeneous structure $\mathcal{M}$—the question of the truth of φ is decidable, modulo the fundamental question stated as Problem B.

The variant of Problem B in which we drop the stability requirement is more general than Problem B and remains open. The problem of amalgamation for relational structures reduces to the case of structures A_1, A_2 extending a common substructure $A_\circ$ by a pair of new points $a_1 \in A_1$ and $a_2 \in A_2$, but this problem remains open except in the binary case, where a direct check produces a finite procedure.

We will give a solution to Problem B. Let $\mathcal{M}$ be the hypothetical structure whose set of constraints $\mathcal{C}$ is specified. The rank of $\mathcal{M}$ is bounded by the number of 2-types and can therefore be computed using quantifier elimination. An inconsistent outcome at this point simply means that $\mathcal{M}$ does not exist. So assume the rank of the still hypothetical structure $\mathcal{M}$ is determined as k. For any definable equivalence relation E on $\mathcal{M}^2$ whose definition involves at most $2k$ parameters, we decide similarly whether the quotient is finite, and if it is finite we determine its size. Let μ bound the size of the finite quotients of this type. Then for any formula $\varphi(x, y; Z)$ one can bound the rank and multiplicity of $\varphi(x, y; B)$ as a function of $tp(B)$. Do so for $|B| \leq 2k$. Let ρ be the arity of $\mathcal{M}$.

Lemma 8.1.1. *Let $\mathcal{M}$ be $\aleph_0$-categorical and $\aleph_0$-stable, and coordinatized by degenerate geometries. Then*

1. *For all $a \in \mathcal{M}$ and $A \subseteq B \subseteq \mathcal{M}$, if $rk(a/B) < rk(a/A)$, then for some $b \in B$ we have $rk(a/Ab) < rk(a/A)$.*
2. *For all $a \in \mathcal{M}$ and $A \subseteq \mathcal{M}$, there is $A_1 \subseteq A$ with $rk(a/A_1) = rk(a/A)$, and $|A_1| \leq rk \mathcal{M}$.*

Proof. Evidently it suffices to deal with the first point, and we may suppose that $B - A$ is finite. We will proceed by induction on $rk(B/A)$. Clearly, $rk(B/A) > 0$.

For $b \in B$, if $b \notin acl(A)$ then choose $b' \in acl(b)$ with $rk(b'/A) = 1$, and otherwise $b' = b$. Set $B' = \{b' : b \in B - acl(A)\}$. As the geometries are degenerate, if $rk(a/B') < rk(a/A)$, then there is an element $b \in B$ with $rk(a/Ab') < rk(a/A)$ and this yields the claim. If $rk(a/B') = rk(a/A)$ then $rk(a/B) < rk(a/B')$ and $rk(B'/A) < rk(B/A)$, so induction applies, yielding

$$rk(a/B'b) < rk(a/B')$$

for some $b \in B$. Let $b'_1, \ldots, b'_n$ be a maximal subset of B' which is independent from b over A. We are assuming a is independent from $b'_1, \ldots, b'_n$ over A, but not from $b'_1, \ldots, b'_n, b$. By the degeneracy of the geometries $rk(a/Ab) < rk A$, as desired. ∎

Lemma 8.1.2. *Let $\mathcal{M}$ be stable, finitely homogeneous, for a language of arity ρ. Let $a, b \in \mathcal{M}$, $A_1 \subseteq A \subseteq \mathcal{M}$, with $rk(ab/A_1) = rk(ab/A)$. Then there is a subset $A_2 \subseteq A$ containing A_1 such that*

$$|A_2 - A_1| \leq \rho \cdot Mult(ab/A_1) \text{ and } Mult(ab/A_2) = Mult(ab/A).$$

Proof. We proceed by induction on $Mult(ab/A_1)$. We may suppose that $Mult(ab/A) < Mult(ab/A_1)$. Take two distinct types over A extending the type $tp(ab/A_1)$, and take a set C of size at most ρ over which they are distinct. Working over $A_1 C$, we conclude by induction. ∎

Definition 8.1.3

1. We consider amalgamation problems of the form $(A; b_1, b_2)$, signifying that a finite relational language $\mathcal{M}$ is specified, $A' = Ab_1$ and $A'' = Ab_2$ are specified finite $\mathcal{M}$-structures agreeing on A, and we seek an amalgam $Ab_1 b_2$ which should omit some specified class of forbidden structures C. We are looking for an amalgam in a stable homogeneous structure and it is assumed that the preliminary analysis of k, μ, and so on, has been carried out in advance as described above.

2. The standard amalgamation procedure for such amalgamation problems $(A; b_1, b_2)$ under the specified conditions is the following:

(i) *Find $E_1, E_2 \subseteq A$ with $|E_i| \leq k$ and $rk(b_i/E_i)$ minimized. (For $|E_i|$ of this size, $rk(b_i/E_i)$ has been given a definite meaning.) Set $A_1 = E_1 \cup E_2$.*

(ii) *For $X \subseteq A$ containing A_1, let $\mathcal{A}(X)$ be the set of amalgams of $b_1 A_1$, $b_2 A_1$, and X over A_1 which omit the specified forbidden structures and satisfy*

$$(*_X) \qquad \text{For } Y \subseteq X, \text{ if } |Y| \leq k \text{ then } rk(b_1 b_2/Y) \geq rk(b_1 b_2/A).$$

These amalgams are not required to be compatible with $b_i X$.

(iii) *Check whether $|\mathcal{A}(X)| \leq \mu$ for all $X \subseteq A$ with $A_1 \subseteq X$ such that $|X - A_1| \leq \rho \binom{\mu}{2}$. If not, then the procedure fails (and halts) at this stage.*

(iv) *Check whether for all subsets $X \subseteq Y \subseteq A$ with $A_1 \subseteq X$, such that $|X - A_1| \leq 2k + \rho \cdot \binom{\mu}{2}$ and $|Y - X| \leq 2\rho$, each element of $\mathcal{A}(X)$ extends to an element of $\mathcal{A}(Y)$. If not, fail and halt.*

(v) *At this point, if the procedure has not failed, then $\mathcal{A}(A) \leq \mu$. Run through the possibilities in $\mathcal{A}(A)$; if one extends Ab_1 and Ab_2, the procedure succeeds.*

Lemma 8.1.4. *Let C be a finite set of constraints (forbidden structures) for the finite relational language $\mathcal{M}$ of arity ρ, all of size at most N. Let k, μ be the invariants associated to a hypothetical stable homogeneous $\mathcal{M}$-structure $\mathcal{M}$ with constraints C, that is the rank and a bound on the sizes of finite quotients of $\mathcal{M}^2$ by equivalence relations definable from $2k$ parameters, computed according to the canonical quantifier elimination procedure from C.*

1. *If there is, in fact, a stable homogeneous $\mathcal{M}$-structure with finite substructures exactly those omitting C, then the standard amalgamation procedure will succeed for any appropriate data $(A; b_1, b_2)$.*

2. *If the standard amalgamation procedure fails for (A, Ab_1, Ab_2), then there is $A' \subseteq A$ of order at most $2k + \rho \cdot \binom{\mu}{2} + \mu \cdot \max(\rho, N)$ for which it fails.*

Proof. The first point has essentially been dealt with in the previous lemmas, modulo the basic properties of independence. For the second, a failure at stage (iii) or (iv) produces a corresponding subset of size at most $2k + \rho \cdot \binom{\mu}{2} + 2\rho$ over which the procedure fails. If the procedure continues successfully to the final step, then $|\mathcal{A}(X)| \leq \mu$ for any X containing A_1. Fix a subset A' of A containing A_1 such that any two possible amalgams differ on $A'b_1b_2$, and $|\mathcal{A}(A')|$ is as large as possible. We may take $|A'| \leq 2k + \rho\binom{\mu}{2}$. For Y containing A' with $|Y - A'| \leq \rho$ each element of $\mathcal{A}(A')$ extends uniquely to $\mathcal{A}(Y)$. With step (iv) this gives a unique extension satisfying the definition of $\mathcal{A}(A)$ apart from the omission of $\mathcal{C}$. Those which omit the forbidden substructures are incompatible with Ab_1 or Ab_2. Thus μ sets of size N or ρ suffice to eliminate all potential solutions to the standard amalgamation procedure, over A'. ∎

Proposition 8.1.5. *Problem B is decidable. Hence Problem A is decidable.*

Proof. Compute the putative rank k and the invariant μ. Attempt the standard amalgamation procedure for all $(A; b_1, b_2)$ with $|A|$ satisfying the bound of the previous lemma. If this fails then the desired structure does not exist. If it succeeds, then there is at least a homogeneous structure $\mathcal{M}$ corresponding to the specified constraints. Furthermore, the quantifier elimination procedure used is correct for $\mathcal{M}$, and so, in particular, its rank has been correctly computed, and it is stable. ∎

8.2 EFFECTIVITY

We continue in the spirit of quasifinite axiomatizability and Ziegler's Conjecture, with attention to issues of effectivity. Recall the notion of a skeletal type and skeletal language L_{sk} from §4.2. From the results in §4.5 we may derive

Lemma 8.2.1. *With the language L and skeletal language L_{sk} fixed, there is a finite set $\mathbf{X}_0(L, L_{\mathrm{sk}})$ of pseudo-characteristic sentences such that*

1. *If $\mathcal{M}$ is a Lie coordinatized L-structure with full skeleton $\mathcal{M}_{\mathrm{sk}}$, then some pseudo-characteristic sentence χ is true in $\mathcal{M}$.*
2. *With $\mathcal{M}$, χ as in (1), every proper model of χ is isomorphic to an envelope of $\mathcal{M}$.*
3. *$\mathbf{X}_0$ is recursive as a function of L and L_{sk}.*

The prefix *pseudo* is called for as no claim is made that *all* of these formulas actually have models. This is the price to be paid, initially, for requiring effectivity.

Proof. This is proved in Proposition 4.4.3 with a potentially infinite set $\mathbf{X}_0$. The finiteness (without regard to effectivity) is in Proposition 4.5.1, by compactness. Paying attention to the effective (and explicit) axiomatizability of the class of structures with the given full skeleton, the effectivity follows from the same argument (via an unlimited search until a proof of a suitable disjunction is found). ∎

Evidently, this is not satisfactory, and we wish to prune off the bogus characteristic sentences, preferably carrying along some side information about dimensions as well, as in the following definition.

Definition 8.2.2. *Assume L and L_{sk} are given.*

1. *A skeletal specification Δ for L_{sk} consists of a skeletal type augmented by dimension specifications for each of the geometries of the forms: "$= n$"; "$\geq n$"; or "$= \infty$," where n stands for a specified finite number (≥ 0 is acceptable, of course). The specification is* complete *if "$\geq n$" does not occur.*
2. *If Δ is a skeletal specification, then $\mathbf{X}_1(L, L_{sk}, \Delta)$ is the set of sentences from $\mathbf{X}_0(L, L_{sk})$ that have a model $\mathcal{M}$ with full skeleton satisfying the specification Δ.*
3. *If Δ is a skeletal specification, then Δ^{∞} denotes its* most general completion*: each specification $\geq n$ is replaced by the specification $= \infty$.*

By definition, Lemma 8.2.1 holds in a sharper form for $\mathbf{X}_1(L, L_{sk}, \Delta)$. We claim further:

Proposition 8.2.3. $\mathbf{X}_1$ *is effectively computable, considered as a function of L, L_{sk}, and Δ.*

This requires substantial argument. We will use induction on the height of the Lie coordinatization. The remainder of this section is devoted to that argument. In particular, L, L_{sk}, and Δ are given. However, we first make some reductions.

First reduction

We replace Δ by Δ^{∞} (so that the characteristic sentences become complete, modulo the underlying theory).

To justify this reduction, note that for any Δ, $\mathbf{X}_0 \backslash \mathbf{X}_1$ is in any case recursively enumerable since it consists of sentences which are inconsistent with the base theory. The problem is to enumerate $\mathbf{X}_1$ effectively. However, each formula φ in $\mathbf{X}_1(L, L_{sk}, \Delta)$ is derivable from another in $\mathbf{X}(L, L_{sk}, \Delta')$ with Δ' complete (working always modulo a background theory). It suffices to handle all the Δ' (uniformly), and as $\Delta' = \Delta'^{\infty}$ the first reduction is accomplished.

Second reduction

We assume that $\mathcal{M}$ is nonmultidimensional and has no "naked" vector spaces.

The point is that these are *conservative extensions*: if a characteristic sentence holds in some $\mathcal{M}$, then that structure can be expanded to a nonmultidimensional one in which, furthermore, every vector space comes equipped with an isomorphism to its definable dual. Compare §5.3. If we can recognize the characteristic sentences in this context, then we can find one that implies the original one (and find the derivation as well). This reduction changes the skeletal type, in an effective way.

Note that if we happen to be interested only in the stable category, at this point the proof leaves that category in any case.

To take advantage of the nonmultidimensionality it is convenient to relax the notion of skeleton, allowing the bottom level to consist of finitely many orthogonal Lie geometries sitting side by side. At higher levels we may restrict ourselves to finite covers and affine covers, with the dual affine part present and covering a self-dual linear geometry lying at the bottom.

As the first level presents no problems, we have only to deal with the addition of subsequent levels, in other words with finite or affine covers. The problem is the following. If $\mathcal{M}$ is the given (hypothetical) structure, and $\mathcal{M}^-$ is the structure obtained from $\mathcal{M}$ by stripping off the top level, then assuming that we can effectively determine what the possibilities for $\mathcal{M}^-$ are, we must determine what the possibilities for $\mathcal{M}$ are. Actually, the emphasis at the outset is on pseudo-characteristic sentences, which while possibly contradictory have at least the virtue of actually existing, rather than the more nebulous $\mathcal{M}$ and $\mathcal{M}^-$, which may not in fact exist. Still the criterion that a pseudo-characteristic sentence χ be acceptable (relative to a given specification Δ) is that there should be an associated χ^- already known to be acceptable, and hence associated with a structure $\mathcal{M}^-$, such that χ^- "says" (or rather implies) that $\mathcal{M}^-$ has a covering of the appropriate type, with the property χ. So we may concern ourselves here with a reduction of the properties of a hypothetical $\mathcal{M}$ to those of a real $\mathcal{M}^-$.

The Case of a Finite Cover

We have $\mathcal{M}^-$, or equivalently a characteristic sentence χ^- for it (which is complete when supplemented by the appropriate background theory including the relevant Δ^- extracted from Δ). We have also a characteristic sentence χ putatively describing a finite cover $\mathcal{M}$ of $\mathcal{M}^-$. Here the details of the construction of these sentences, in the proof of quasifinite axiomatizability, become important. The point is that χ gives a highly overdetermined recipe for the explicit determination of all structure on $\mathcal{M}$, proceeding inductively along an Ahlbrandt–Ziegler enumeration; if one begins with the structure $\mathcal{M}$, one of course writes down the facts in $\mathcal{M}$, but to capture all possible χ is a matter of writing down all conceivable recipes, most of which presumably have internal contradictions. The problem is to detect these contradictions effectively by confronting χ with $\mathcal{M}^-$.

Let K be a bound for the various numbers occurring in the proof of Proposition 4.4.3, say $K = 2k + \max(k^*, k^{**}) + 1$. Let d be the Löwenheim-Skolem number associated with K in $\mathcal{M}^-$; i.e., any K elements of $\mathcal{M}^-$ lie in a d-dimensional envelope in $\mathcal{M}^-$ (effectively computable, by Lemma 5.2.7). Test χ by testing the satisfiability of χ in a finite cover of such a d-dimensional envelope (by a search through all possibilities). Here we should emphasize that χ is of the specific form given in the proof of Proposition 4.4.3, so that if true in some $\mathcal{M}$ it would pass to this particular envelope.

Conversely, if χ passes this test, we claim that the construction of $\mathcal{M}$ according to χ succeeds. Running over an Ahlbrandt–Ziegler enumeration of $\mathcal{M}^-$, at each stage we have covered certain elements of $\mathcal{M}^-$ by appropriate finite sets with additional structure, and have the task of covering one more element a of $\mathcal{M}^-$ by a finite set, and specifying its atomic type over everything so far.

Look for a formula $\theta(x, \mathbf{y})$, where x refers to the elements of the fiber being added, and $\mathbf{y}$ (of length at most k) refers to k previously constructed elements, with the following properties:

1. χ implies that such an x exists (more on this momentarily),
2. the multiplicity of x over everything so far is minimized, according to θ.

Let us consider (1) more carefully. We require previously constructed elements $\mathbf{z}$ and a valid atomic formula $\rho(\mathbf{y}, \mathbf{z})$, such that

$$\chi \quad \Longrightarrow \quad \forall \mathbf{y}, \mathbf{z}[\rho(\mathbf{y}, \mathbf{z}) \Longrightarrow \exists x \theta(x, \mathbf{y})].$$

We then hope to see the following:

3. For all $\mathbf{y}'$, there are $\mathbf{z}'$ such that χ together with the atomic type of $\mathbf{y}, \mathbf{y}', \mathbf{z}'$ will imply the atomic type of $x, \mathbf{y}'$.
4. After adding x as specified, the universal part of χ holds.

If any of these hopes are disappointed, then the failure is witnessed by at most K elements and hence is also visible in the envelope with dimensions d.

One of the simplifying features in this case is that "everything is algebraic." In the case of affine covers, the behavior of algebraic closure in the hypothetical cover is one of the sticking points. For this the affine dual is helpful.

The Case of Affine Covers

We first shift the notation slightly. We may suppose that the dual-affine part of the cover is absorbed into $\mathcal{M}^-$, since it is a finite cover of a linear geometry in $\mathcal{M}^-$—just apply the previous case.

The following remark may be useful as motivation. Since the dual affine part is present in $\mathcal{M}^-$, $\mathcal{M}$ is rigid over $\mathcal{M}^-$; that is, the extension is canonical, but not definable. Questions of multiplicity do not arise, and the question of existence of $\mathcal{M}$ is transformed into a different question: does the *canonical $\mathcal{M}$* have the posited property χ? It will suffice to show that this can be expressed in $\mathcal{M}^-$.

We fix the following notation: $V = V^*$ is the linear geometry in $\mathcal{M}^-$; A^* is an affine cover (with components A_t^*, each a finite cover of V^*); A is the affine cover, in $\mathcal{M}$ but not in $\mathcal{M}^-$, with components A_t dual to A_t^*.

The elements $a \in A_t$ will be identified with hyperplanes in A_t^* which project bijectively onto V^*. From this point of view, the problem is one of elimination of a second-order quantifier (for such hyperplanes) from the language of $\mathcal{M}^-$.

Lemma 8.2.4. *Let $\mathcal{M}_0$ be the reduct of $\mathcal{M}$ including all structure on $\mathcal{M}^-$ (which we take to include the affine duals A^*) as well as the geometrical structure on A: affine space structure of A_t over V, and duality with A_t^*. Then this is the full structure on $\mathcal{M}$ (all 0-definable relations remain 0-definable).*

Proof. It suffices to show that if two tuples a, b have the same types in the reduct, then they have the same types. Take an envelope E containing them and view the affine elements in a, b as predicates (for hyperplanes). These predicates are conjugate under the automorphism group of E^- (the top layer is stripped off) by assumption, and any such automorphism extends to one of E. Thus a, b have the same type in the full language. ∎

Lemma 8.2.5. *Let $\mathcal{M}^-$ be a countable (or hyperfinite) Lie coordinatizable structure with distinguished sorts T, V, V^*, A^* with the usual properties; e.g., A^* is a T-parametrized family of affine dual covers of V^* (or more generally V_t^*), possibly with additional parameters fixed. Then there is a cover by an affine sort $A = \bigcup_T A_t$ compatible with the affine duals A_t^*, in the geometric language of the previous lemma, and its theory is uniquely determined.*

Proof. For the existence, we may assume $\mathcal{M}^-$ is nonmultidimensional (as we have been, in any case) and does not have quadratic geometries (it suffices to adjoin some parameters). The issue of orientability falls away and $\mathcal{M}$ can be thought of as nonstandard-finite. In this case, existence follows from the finite case: adjoin all internal linear sections for the maps $A_t^* \to V_t^*$ in a nonstandard universe, and this is locally Lie, hence Lie.

For the uniqueness of the theory, fix a formula, and shrink a given affine expansion to a finite envelope large enough to test the truth of the formula; at the finite level the expansion is completely canonical; hence the answer is determined. ∎

Lemma 8.2.6. *In the context of the previous lemma, the theory of the affine expansion $\mathcal{M}$ can be computed from the theory of $\mathcal{M}^-$.*

Proof. Follow the line of the previous argument. One needs to determine the theory of a finite envelope $\mathcal{M}_d$. This is the canonical expansion of a finite envelope $\mathcal{M}_d^-$. Its theory can be determined by inspection.   ∎

8.3 DIMENSION QUANTIFIERS

In this section we consider enhancements of first order logic expressing numerical properties of geometries in large finite (or nonstandard-finite) structures. That some such expansion is necessary to carry through the analysis of Lie coordinatization in a definable and effective way is made clear by the following example given in [HrBa].

Let V be a finite dimensional vector space over a finite field, and let m, n be distinct nonnegative integers. Let $V_{m,n}^3$ be a free cover of the cartesian cube V^3 by finite sets of sizes m or n; the triple (v_1, v_2, v_3) will be covered by a set of size m if $v_3 = v_1 + v_2$, and by a set of size n otherwise. Let $\mathcal{M}(m, n)$ be the reduct of $V_{m,n}^3$ in which the vector space structure of V is forgotten. We can view this as having sorts V and V^3 in addition to the covering M, with the covering map $\pi : M \to V^3$ and the projections from V^3 to V. The collection $\mathcal{M}(m, n)$ should be thought of as a uniform family of examples, but the recovery of the vector space structure from the covering is nonuniform with respect to first order logic. In the usual approach to effectivity, one sorts out all the structures under consideration into finitely many classes, each axiomatizable in first order logic. We propose to follow much the same route here, after augmenting the logic to allow us to decode numerical information of the type used here: note that it is not necessary to know the value of m and n, but only which is larger (or actually, with a little more care, that they are different). This will be done using a *dimension comparison quantifier* to be introduced shortly.

The specific quantifier introduced in [HrBa] in its "most general form" is actually too general, as we will now indicate. The simplest way to add the desired numerical quantifier would be with a *less than* quantifier "$<$". Applied to two formulas φ, φ' involving the variable x, and possibly other free variables, the formula $<_x(\varphi; \varphi')$ would represent the formula: *the cardinality of the set defined by φ is less than the cardinality of the set defined by φ'*; as usual, variables other than x which are free in φ or φ' remain free in the quantified expression. The problem with this is that it encodes undecidable problems—namely, any diophantine problem over $\mathbb{Z}$—into the basic properties of structures with a bounded number of 4-types (in fact, directly into a multisorted theory of pure equality). A polynomial equation $p(\mathbf{x}) = 0$ may be encoded as an equation with nonnegative coefficients $p_1(\mathbf{x}) = p_2(\mathbf{x})$, and after interpreting multiplication as cartesian product and sum as disjoint union, the solvability of such an equation is equivalent to the existence of a model $\mathcal{M}$ of the theory of equality with a number of sorts equal to the number of variables **x**, satisfying one additional sentence involving the cardinality quantifier (which expresses the stated equality). We require a less expressive logic, for which we can determine effectively whether a Lie coordinatizable structure with a specified number of 4-types exists, having any specified property expressible in the logic.

Strictly speaking, we will make use of three enhancements of first order logic: a finite set of *fully embedded geometry* quantifiers G_t, a *dimension comparison* quantifer $D<$, and the standard quantifier $\exists^\infty$—there are infinitely many. The second has a natural model only in finite structures, where the third encounters a frosty reception, so we will have to pay some attention to *weak* (i.e., not canonical) interpretations of the logic as well. We will need completeness and compactness theorems for various combinations of these notions, in a limited context (essentially the context of Lie coordinatizable structures). Our specification of intended interpretations below will be less useful from a technical point of view than the axioms specified subsequently, determining the notion of a "weak" interpretation.

Definition 8.3.1

1. *A type t (of geometry) is one of the following:* (i) *set;* (ii) *linear;* (iii) *orthogonal$^-$;* (iv) *orthogonal$^+$;* (v) *symplectic;* (vi) *unitary. For each type t, the quantifier G_t has the syntax of an ordinary quantifier: if φ is a formula, then $G_t x \varphi$ is also a formula, with x bound by G_t. The intended interpretation in a model $\mathcal{M}$ is that the subset of $\mathcal{M}$ defined by $\varphi(x)$ is a fully embedded geometry of type t. The distinction between the two types of orthogonal geometry has a clear meaning only in the finite case, but will be carried along formally in all cases (in other words, the Witt defect is included in the type). As usual, variables other than x which are free in φ remain free in $G_t x \varphi$, and have the effect of auxiliary parameters.*

2. *The lesser dimension quantifier $D<$ acts on pairs of formulas φ, φ' to produce a new formula $Dx(\varphi < \varphi')$. The intended meaning in a structure $\mathcal{M}$ is that*

   (i) *φ and φ' define fully embedded canonical projective geometries J, J' of the same type; and*
   (ii) *$\dim J < \dim J'$.*

Evidently, (i) is already expressible using the G_t.

3. *The quantifier $\exists^\infty$ is the usual quantifier "there exist infinitely many." It may also have nonstandard interpretations in finite models, essentially of the form "there exist a lot."*

4. *The logics $\mathcal{M}^G$, $\mathcal{M}^D$, $\mathcal{M}^{D\infty}$ are obtained syntactically by augmenting first order logic by, respectively: all the G_t; all the G_t, and $D<$; all the G_t, $D<$, and $\exists^\infty$. In each case the logic is taken to be closed under iterated applications of all the operations.*

Context. Our basic context will consist of a fixed finite language together with a specified bound k on the number of 4-types; the latter is formalized by a theory which we denote $B4(k)$; more exactly $B4(L, k)$ where L is the logic in use. (The richer the language, the more powerful this theory becomes.) In finite models with at most k 4-types, the language $\mathcal{M}^D$ has a canonical interpretation. We write $C_4(L, k)$ for the class of finite L-structures with at most k 4-types.

Proposition 8.3.2 (Effective Coordinatizability). *There is a computable function $b(L, k)$ such that with the language L and the bound k fixed, every $M \in C_4(L, k)$ has a Lie coordinatization via formulas in $\mathcal{M}^D$ of total length at most $b = b(L, k)$.*

Proof. Both the boundedness and the effectivity are at issue.

For the boundedness, we use a modified compactness argument. Suppose toward a contradiction that $\mathcal{M}_n \in C_4(L, k)$ has minimal coordinatization of total length at least n, for each n. Without loss of generality these all involve the same skeleton (but the actual definitions of the geometries vary erratically). Consider the first order structure $\mathcal{M}_n^*$ obtained by adjoining predicates to $\mathcal{M}_n$ for all formulas in $\mathcal{M}^D$, as well as predicates giving the appropriate coordinatization. (Note that as $\mathcal{M}_n$ is finite, this does not affect definability in the individual structures, but does change the collection of uniformly definable relations as n varies.) Pass to an ultraproduct $\mathcal{M}_\infty^*$. This is weakly Lie coordinatized. Let $\mathcal{M}_\infty$ be the reduct of $\mathcal{M}_\infty^*$ to $\mathcal{M}^D$ (or rather the first-order language used to encode $\mathcal{M}^D$ in the $\mathcal{M}_n$). By the *theorem on reducts* this is also Lie coordinatizable, definably. One would like to say that this "property" is inherited by the $\mathcal{M}_n$. By the proof of quasifinite axiomatizability, there is a sentence which characterizes the *envelopes* in $\mathcal{M}_\infty$, for models whose dimensions are true (constant over geometries parametrized by realizations of the same type). Use of $\mathcal{M}^D$-definable predicates ensures that the $\mathcal{M}_n$ have true dimensions in this sense, and hence are envelopes. In particular they are Lie coordinatizable uniformly, contradicting their choice.

Now we turn to the effectivity of $b(k)$. There is a set of formulas in the language $\mathcal{M}^D$ which is adequate for the Lie coordinatization of any structure in our class. We wish to argue that this is a first order property and is a consequence of an explicitly known theory, and then to conclude via the completeness theorem.

As a base theory one may take a first order theory in which all $\mathcal{M}^D$ formulas occur as atomic predicates, and their definitions—to the extent that they have definitions—are included as axioms. To a very large extent the $\mathcal{M}^D$ formulas do have first order definitions, since it is possible to say in a first order way what the dimension is when it is finite. Thus we may include in the axioms: if a given dimension is finite (i.e., specified explicitly), then it is *formally* less than another if and only if it is, in fact, less than that other. These axioms leave open what happens when the dimensions are infinite. (In general, it is a good idea to require that "less than" be transitive, but this is not yet relevant.)

Now for $b \geq b(k)$, there is a finite disjunction of potential Lie coordinatizations, and a corresponding collection of characteristic sentences (in the sense of the previous section) for which, in fact, one of the coordinatizations works within every structure of our class, and one of the corresponding characteristic sentences is valid. This is a first order sentence. Furthermore, whenever the appropriate characteristic sentence is valid, the corresponding Lie coordinatization is, in fact, a valid Lie coordinatization. This is the delicate point: to verify that a potential Lie coordinatization is in fact valid, it is necessary to have complete control over definability; for example, one must know that if no vector space structure is specified on a set, then it has no definable vector space structure. The characteristic sentences give this kind of control.

Accordingly, one can search for a provable first order sentence of the desired form, and when it is found then one has found an effective bound on $b(k)$ (we are not concerned here with the minimum value of $b(k)$). ∎

Now we will develop a completeness theorem for $\mathcal{M}^D$ and use it to produce more explicit results on effectivity.

Definition 8.3.3. $TF4_k$ *is the following axiom system, whose models are called* weak models *for* $\mathcal{M}^D$.

1. *Background axioms as in the preceding proof: predicates correspond to all formulas of $\mathcal{M}^D$ and the axioms force "formal less than" to mean "less than" when at least one of the numbers is finite.*
2. *There are at most k pairwise contradictory formulas in 4 variables.*
3. *For the quantifiers G_t, assert that when they hold then the corresponding geometry is embedded and stably embedded.*
4. *Some group of formulas of total length less than $b(k)$ (from the preceding lemma) forms a Lie coordinatization. Use the quantifiers G_t here.*
5. *Transitivity of the relation "$\dim(J) < \dim(J')$." (Supplementing (1) above.)*
6. *If the definable set D is not a canonical Lie geometry, then some formula of length at most $b'(k)$ shows that it is not. Here $b'(k)$ is also effective; failure involves failure of primitivity, rank bigger than 1, or a richer Lie structure than the one specified is definable. In all cases there is a definable predicate that shows this. The bound $b'(k)$ can be found in the same way as $b(k)$.*

Proposition 8.3.4. *Let φ be a sentence in $\mathcal{M}^D$ which is consistent with the axioms given above. Then φ has a finite model with at most k 4-types.*

Proof. Begin with a weak model, which will be Lie coordinatized. Note that if it is finite, then it already has all required properties as they are expressed by the theory in this case. Otherwise, shrink it (i.e., take an envelope), preserving the truth of φ by keeping infinite dimensions large. Note that the formal less than relation on the infinite dimensions determines a linear ordering of finite length and hence can be respected by the shrinking process. (Note that the position in this sequence of a given infinite dimension is part of the type of the associated parameter to begin with.) ∎

Corollary 8.3.5. $TF4_k$ *is decidable, uniformly in k.*

Proposition 8.3.6. *Extend the logic by the quantifier $\exists^\infty$ to get $\mathcal{M}^{D,\infty}$. The theory remains decidable.*

Proof. One must extend the axiom system to get a suitable notion of weak model, then convert each weak model into one in which all sets whose size is formally not infinite become sets which are in fact finite. To avoid pathology (or paying more attention over the formalization) one may suppose all structures contain at least two elements.

The axioms are as follows. We use the term "finite" here for "definable and formally finite" rather than "of specified size."

1. $\exists^\infty$ implies the existence of arbitrarily many (the conclusion is a first order scheme).
2. If $\exists^\infty x \exists y \varphi(x, y)$, then either $\exists y \exists^\infty x \varphi(x, y)$ or $\exists^\infty y \exists x \varphi(x, y)$. In other words, the image of an infinite set under a finite-to-one function is infinite.
3. A definable subset of a definable finite set is finite.
4. Given two embedded, stably embedded geometries, one of which is formally infinite, and the other having dimension at least as large, then the second geometry is also formally infinite. (This relates $\exists^\infty$ and the dimension quantifier.)

Note that (2) implies that a finite union of finite sets is finite.

The problem now is to take a formula φ which has a weak model and give it a model in which all sets asserted to be of finite size are in fact of finite size. We may assume that φ specifies a coordinatization, and using $(2, 3)$ we may also assume that the only sets whose finitude or infinitude are asserted are subsets of canonical projective geometries (possibly degenerate), and in view of the nature of definability in such geometries, we reduce further to the finitude or infinitude of the geometry itself. So the problem is to shrink geometries which are asserted to be of finite size to ones which are finite, while leaving alone those asserted to be infinite, and preserving both the order relationships (for which (4) is clearly essential, and largely sufficient) and the other (essentially first order) properties asserted by φ. Note that axiom (1) is not required to "do" a great deal; but it guarantees that unmitigated sloth is an adequate treatment of the infinite case.

For all of this to make sense, one thing is necessary: the formally finite and the formally infinite canonical projective geometries should be orthogonal (otherwise, there is no appropriate dimension function to begin with). This is guaranteed by (2, 3).   ∎

8.4 RECAPITULATION AND FURTHER REMARKS

We return very briefly to the survey given in the Introduction. The theory of envelopes was summarized in Theorem 1 and in terms of finite structures in Theorem 6, the latter incorporating the numerical estimates of §5.2 and some effectivity. The families referred to in Theorem 6 are determined by a specific type of Lie coordinatization in the language $\mathcal{M}^D$ as well as a definite characteristic sentence. Evidently, the truth of a sentence can be determined in polynomial time. Part (5) of Theorem 6 is dealt with in §5.2, as far as sizes go, and the construction is given by the characteristic sentence.

Theorem 2 gave six conditions equivalent to Lie coordinatizability. The first five conditions were dealt with by the end of §3.5; this is discussed at the beginning of that section. In particular, to get from Lie coordinatizability to smooth approximability one uses the theory of envelopes, notably §3.2. The converse direction was the subject of §3.5. For the validity of the last condition, use Lemma 5.2.7 and the estimate on the sizes of envelopes.

Theorem 3 is the theory of reducts, given in §7.5. Theorem 5 summarizes the effectivity results of §§8.1–8.3. Theorem 7 has been dealt with in §7.5.

We recall one problem mentioned in [HrBa]: are envelopes "constructible" in time polynomial in the dimension function? As noted there, the underlying sets are, in fact, too large to be constructed in polynomial time, but the problem has a sensible interpretation: the underlying set can be treated as known, and one can ask whether the basic relations on it can be recognized in polynomial time (for example, think of the basic case in which the envelope is simply a geometry of specified dimension). This problem has model theoretic content. The proof of quasifinite axiomatizability is based on a 1-way version of "back-and-forth" which may be called "carefully forth." We do not know how to give this proof in a "back-and-forth" format, and it seems that the polynomial time problem involves difficulties of the type which have been successfully eluded here.

8.4.1 The role of finite simple groups

In view of the special status of the classification of the finite simple groups it seems useful both to clarify the dependence of the present paper on that result, and to consider the possibilities for eliminating that dependence, and arguing in the opposite direction.

The work carried out here can be viewed as a chapter within model theory which is dependent in part on the classification of the finite simple groups for its motivation, but which in terms of its content is largely independent of that classification both logically and methodologically.

For example, Theorem 7 as we have stated it is independent from that classification. Similarly, the proof of Theorem 6 really involves Lie coordinatizable structures, and as such does not involve the classification of the finite simple groups, which is invoked at the end, via Theorem 2, to give the present statement of that result. As far as Theorem 2 is concerned, we combine the primitive case from [KLM], which may be taken here as a "black box," with independent model theoretic methods.

However, the proof of [KLM] is strongly dependent on the classification of the finite simple groups. Theorem 7 offers an array of model theoretic properties which can be taken as defining a certain portion of the theory provided by the classification of the finite simple groups. No such model theoretic version is known for the whole classification, and for that matter we are not aware of any other comparable portion of the classification that can be expressed in model theoretic terms. Initially one might try to assume Theorem 2-Characterizations (3) (i.e., 2 (6) with an arbitrary function), so that one has (LC1) and (LC2), and ask whether one can prove (LC3–LC9) directly and not inductively. The combinatorial flavor of the properties (LC4–LC9) suggests that this may not be an unreasonable endeavor.

This issue was raised in [HrBa] and remains both open and of considerable interest. It was noted there that the results on sizes of definable sets can be reversed to give a definition of rank and indpendence in purely combinatorial terms, that is in terms of asymptotic sizes of sets. In particular the properties (LC4) and (LC5) then become cleanly combinatorial. Property (LC4) becomes the statement that model-theoretically independent subsets of a single type over an algebraically closed base are statistically independent (giving unexpected support for the old term: "independence theorem"). We give a direct proof of this below. This proof is closely analogous to the proof of (LC4) from finite S1-rank given in [HrS1]], but it emerged only on following up a suggestion of L. Babai regarding the similarity of the desired result with Szemeredi's regularity lemma, a similarity which will not be pursued here. The next challenge, accordingly, would be a direct proof of (LC5).

In the following, we work with the extension of first order logic by cardinality quantifiers, allowing us to assert that one definable set is smaller than another, and also allowing cardinality comparisons of the form $m|D| < n|D'|$, via some definable encoding of disjoint unions. This could be recast more generally in a context where one has a definable probability measure on the definable sets. Indeed, the general relation between simplicity and the existence of such probability measures remains to be clarified.

Let $\mathcal{M}$ be a nonstandard member (e.g., an ultraproduct) of a family of finite structures, where cardinality quantifiers receive their canonical interpretations in finite structures, and the corresponding nonstandard interpretations in the ultraproduct. Call a definable set D *small* if $|D|/|M|$ is infinitesimal, where M is the underlying set of $\mathcal{M}$.

Lemma 8.4.1. *If D forks over $\emptyset$ then D is small.*

Proof. We may suppose that D divides over $\emptyset$; that is, D has an an arbitrarily large indiscernible set $\{D_i\}$ of conjugates which is k-inconsistent for some fixed k. It follows by induction on k that D is small; more exactly (for the sake of the induction) that $|D_i|/|\bigcup D_i|$ goes to 0 as the size r of the set of conjugates increases. If $k = 1$, then these sets are empty, and for $k > 1$ we may consider for each i the $(k-1)$-inconsistent family $\{D_i \cap D_j\}$ for $j \neq i$. Then by induction $|D_i \cap D_j|/|D_i|$ goes to 0 as r increases, so the cardinality of a union of length n of conjugates D_i is of the order of $n|D_i|$, as long as $\binom{n}{2}|D_i \cap D_j|/|D_i|$ is negligible. ∎

Lemma 8.4.2. *Suppose that $\mathcal{M}$ is a nonstandard member of a family of finite structures that realize boundedly many 4-types. Let p_1, p_2, p_3 be 1-types, and let p_{12}, p_{13}, p_{23} be 2-types projecting onto the corresponding 1-types appropriately. Then there is a formula $\varphi(x, y)$ such that $\varphi(a_1, a_2)$ holds if and only if $\{y : p_{13}(a_1, y)\&p_{23}(a_2, y)\}$ is small, and this formula is stable, and is even an equation in the sense of Srour [PS].*

Proof. The set $D = \{y : p_{13}(a_1, y)\&p_{23}(a_2, y)\}$ is definable from two parameters and can take on only a finite number of cardinalities in $\mathcal{M}$ (as this holds, with a bound, in the family of finite structures associated with $\mathcal{M}$). Hence φ can be defined. Now we must show that if (a_i, b_i) is an indiscernible sequence, and $\varphi(a_i, b_j)$ holds for $i < j$, then $\varphi(a_i, b_i)$ holds for all i. Let $D_i = \{y : p_{13}(a_i, y)\&p_{23}(b_i, y)\}$. Then by assumption $|D_i \cap D_j|/|M|$ is infinitesimal for $i \neq j$. As in the previous argument, if $|D_i|$ is not small relative to $|M|$, then $|D_i|$ is small relative to $\bigcup D_i$, and hence also relative to $|M|$, a contradiction. ∎

Proposition 8.4.3. *With the hypotheses of the preceding lemma, suppose that there is no finite 0-definable equivalence relation splitting p_i ($i = 1$, 2, or 3), and that p_{ij} is not small relative to $\mathcal{M}^2$ for $i, j = 1, 2; 1, 3; 2, 3$. Let P_{123} be the set of triples $(a_1, a_2, a_3) \in \mathcal{M}^3$ such that $\mathcal{M} \models p_{ij}(a_i, a_j)$ for each pair $i, j = 1, 2; 1, 3; 2, 3$. Then P_{123} is not small relative to $\mathcal{M}^3$, and, in particular, is nonempty.*

Proof. We use similar notations P_i, P_{ij} for the loci of the given types.

We may compute the number of triples (a_1, a_2, a_3) satisfying $p_{13}(a_1, a_3)$ and $p_{23}(a_2, a_3)$ by first choosing a_3 in $|P_3|$ ways, then choosing a_i for $i = 2, 3$ in $|P_{i3}|/|P_3|$ ways; this yields $|P_{13}||P_{23}|/|P_3|$, which is not small relative to

$|M^3|$. It follows that for some a_1 satisfying p_1, the number of a_2 for which $\neg\varphi(a_1, a_2)$ holds is not small relative to $|M|$, and hence the formula $\neg\varphi(a_1, x)$ does not fork over $\emptyset$. Hence $\neg\varphi(a_1, a_2)$ holds for some pair (a_1, a_2) which is φ-independent in the sense of local stability theory. Then by stability and our hypothesis on p_1, p_2, $\neg\varphi(a_1, a_2)$ holds for all such independent pairs. Similarly, we can choose φ-independent (a_1, a_2) satisfying p_{12}. So all solutions to p_{12} satisfy $\neg\varphi$, and the claim follows.   ∎

We have not touched on the other directions for further research which were already mentioned in [HrBa]. As far as the diagonal theory envisaged there is concerned, the completion, or near-completion, of the foundations of geometric simplicity theory ought to be helpful in this connection.

References

[AZ1] G. Ahlbrandt and M. Ziegler, Quasifinitely axiomatizable totally categorical theories, *Annals of Pure and Applied Logic* **30** (1986), 63–82.

[AZ2] G. Ahlbrandt and M. Ziegler, What's so special about $(\mathbb{Z}/4\mathbb{Z})^{\omega}$?, *Archive Math. Logic* **31** (1991), 115–132.

[BeLe] G. Bergman and H. Lenstra, Subgroups close to normal subgroups, *J. Algebra* **127** (1989), 80–97.

[Bu] S. Buechler, *Essential Stability Theory*, Perspectives in Mathematical Logic, Springer-Verlag, New York, 1996, xi+355 pp. ISBN 3-540-61011-1.

[CaP] P. Cameron, Finite permutation groups and finite simple groups, *Bull. London Math. Society* **13** (1981), 1–22.

[CaO] P. Cameron, *Oligomorphic Permutation Groups*, London Math. Society Lecture Notes Series **152**, Cambridge University Press, Cambridge, England, 1990.

[CaK] P. Cameron and W. Kantor, 2-Transitive and antiflag transitive collineation groups of finite projective spaces, *J. Algebra* **60** (1979), 384–422.

[CaL] R. Carter, *Simple Groups of Lie Type*, Wiley Classics Library, London, 1989.

[Ch] G. Cherlin, Large finite structures with few types, in *Algebraic Model Theory*, eds. B. Hart, A. Lachlan, M. Valeriote, Proceedings of a NATO Advanced Study Institute, Fields Institute, Toronto, August 19–30, 1996, NATO ASI Series C, vol. 496., Kluwer, Dordrecht, 1997.

[CHL] G. Cherlin, L. Harrington, and A. Lachlan, $\aleph_0$-categorical, $\aleph_0$-stable structures, *Annals of Pure and Applied Logic* **18** (1980), 227–270.

[CL] G. Cherlin and A. Lachlan, Stable finitely homogeneous structures, *Transactions American Math. Society* **296** (1986), 815–850.

[CoAt] J. H. Conway, R. T. Curtis, S. P. Norton, R. A. Parker, and R. A. Wilson, *Atlas of Finite Groups; Maximal Subgroups and Ordinary Characters for Simple Groups*, Clarendon Press, Oxford University Press, Oxford, England, 1985.

[EvSI] D. Evans, The small index property for infinite-dimensional classical groups, *J. Algebra* **136** (1991), 248–264.

[Ev] D. Evans, Computation of first cohomology groups of finite covers, *Journal of Algebra* **193** (1997), 214–238.

[EH] D. Evans and E. Hrushovski, On the automorphism groups of finite covers, *Annals of Pure and Applied Logic*, **62** (1993), 83–112.

[FJ] M. Fried and M. Jarden, *Field Arithmetic*, Ergebnisse der Mathematik und ihrer Grenzgebiete, 3. Folge, Band **11**, Springer-Verlag, New York, 1986.

[Ha] L. Harrington, Lachlan's homogeneous structures, lecture, ASL conference,

Orsay (1985).

[Hi] D. G. Higman, Intersection matrices for finite permutation groups, *J. Algebra* **6** (1967), 22–42.

[HoPi] W. Hodges and A. Pillay, Cohomology of structures and some problems of Ahlbrandt and Ziegler, *J. London Mathematical Society* **50** (1994), 1–16.

[HrTC] E. Hrushovski, Totally categorical structures, *Transactions American Math. Society* **313** (1989), 131–159.

[HrS1] E. Hrushovski, Pseudofinite fields and related structures, §§4,6, preprint, 1992, 40 pp.

[HrBa] E. Hrushovski, Finite structures with few types, in *Finite and Infinite Combinatorics in Sets and Logic*, eds. N. W. Sauer, R. E. Woodrow, and B. Sands, NATO ASI Series C vol. 411, Kluwer, Dordrecht, 1993.

[JP] W. Jones and B. Parshall, On the 1-cohomology of finite groups of Lie type, 313–318, in *Proceedings of the Conference on Finite Groups, Utah*, eds. F. Gross and W. Scott, Academic Press, New York, 1976.

[KLM] W. Kantor, M. Liebeck, and H. D. Macpherson, $\aleph_0$-categorical structures smoothly approximable by finite substructures, *Proceedings London Math. Society* **59** (1989), 439–463.

[KiTh] B. Kim, *Simple First Order Theories*, doctoral thesis, University of Notre Dame, Notre Dame, Indiana, 1996.

[Ki] B. Kim, Forking in simple unstable theories, *J. London Math. Society* **57** (1998), 257–267.

[KiP] B. Kim and A. Pillay, Simple theories, *Annals of Pure and Applied Logic* **88** (1997), 149–164.

[KL] J. Knight and A. Lachlan, Shrinking, stretching, and codes for homogeneous structures, in *Classification Theory*, ed. J. Baldwin, Lecture Notes in Mathematics **1292**, Springer, New York, 1985.

[LaPP] A. Lachlan, Two conjectures regarding the stability of ω-categorical theories, *Fundamenta Mathematicæ* **81** (1974), 133–145.

[La] A. Lachlan, On countable stable structures which are homogeneous for a finite relational language, *Israel J. Math.* **49** (1984), 69–153.

[Mp1] H. D. Macpherson, Interpreting groups in ω-categorical structures, *J. Symbolic Logic* **56** (1991), 1317–1324.

[Mp2] H. D. Macpherson, Homogeneous and smoothly approximated structures, in *Algebraic Model Theory*, eds. B. Hart, A. Lachlan, and M. Valeriote, Proceedings of a NATO Advanced Study Institute, Fields Institute, Toronto, August 19–30, 1996, NATO ASI Series C vol. 496. Kluwer, Dordrecht, 1997.

[PiGS] A. Pillay, *Geometrical Stability Theory*, Oxford Logic Guides **32**, Clarendon Press, Oxford, England, 1996, x+361 pp. ISBN 0-19-853437-X.

[PiGr] A. Pillay, Definability and definable groups in simple theories, *J. Symbolic Logic* **63** (1998), 788–796.

[PS] A. Pillay and G. Srour, Closed sets and chain conditions in stable theories, *J. Symbolic Logic* **49** (1984), 1350–1362.

[PoGS] B. Poizat, *Groupes stables*, Nur al-Mantiq wal-Ma'rifa, Villeurbanne, 1987, 215 pp. (illustr.).

[Sch] G. Schlichting, Operationen mit periodischen Stabilisatoren, *Arch. Math. Basel* **34** (1980), 97–99.

[ShS]   S. Shelah, Simple unstable theories, *Annals of Mathematical Logic* **19** (1980), 177–203.

[Sh]   S. Shelah, *Classification Theory*, first edition 1978; revised edition 1990, xxxiv+705 pp.; Elsevier, Amsterdam, The Netherlands.

[Wa]   F. Wagner, Almost invariant families, undated preprint, 4 pp.

[Wi]   E. Witt, Theorie der quadratischen Formen in beliebigen Körpern, *Crelle's Journal* **176** (1937), 31–44, §1.

Index

affine
   see *geometry; isolation*
Ahlbrandt, Gisela: 2, 8, 63, 126
   see also *AZ enumeration*
Amaal: 10, 191
amalgamation
   see *type*
approximable
   smoothly: 5, 12, 61, 149*ff*
   weakly: 5, 12
axiomatizability: 1, 6, 170*ff*
AZ enumeration: 69*f*, 175*f*

canonical
   base: 97
   see also *geometry*
$\aleph_0$-categoricity: 7, 9, 12, 30, 167
Cauchy, Augustin-Louis: 3
cohomology: 48, 81
component, principal: 131
coordinatizable: 4*ff*, 17, 21, 30, 36
coordinatization: 2, 16*ff*, 54
   local: 54*ff*, 183
   weak: 6, 7, 61, 84*ff*, 120, 164
cover: 80
   bilinear: 127*ff*
   finite: 56, 89, 104, 126, 165, 175
   semi-dual: 9, 10, 126*ff*

defect, Witt: 6, 14, 15, 37, 149
dimension
   dimension function: 40, 92
   of Grassmannian: 46
dual
   affine: 8, 13, 27, 45, 68, 126, 138, 176
   linear: 10, 14, 27, 45, 121
   over prime field: 27
   see also *cover, semi-dual*

effectivity: 170*ff*
elimination
   see *quantifier; imaginary elements*
embedding
   canonical: 16, 33, 46*ff*
   full: 16, 29, 32, 36, 74
   stable: 8, 10, 16
   see also *homogeneity; word*
enumeration
   see *AZ enumeration; standard*
envelope: 1, 6, 40*ff*, 124, 149, 168, 173
   equidimensional: 149
   size: 90*ff*, 120
Evans, David: 6, 7, 81, 149

filtered: 107
finite
   basis: 7, 126, 134*ff*
   geometrically: 8, 63, 73
   model property: see *pseudofinite*
free over: 42
Fried, Michael: 11

Galois connection: 3
general position: 7, 167
generic
   equivalence relation: 85, 162
   see also *independence*
geometry
   affine: 8, 16
   basic: 16, 20
   canonical projective: 8, 36*ff*, 79, 89*ff*, 150, 179
   degenerate: 14, 47, 48, 153*ff*
   Lie: 13
   linear: 7, 13
   oriented: 15
   orthogonal: 14, 65, 90

geometry
  pointed: 52
  polar: 14, 15
  projective: 16
  quadratic: 14-15, 45
  reduced: 31
  semiprojective: 16
  simple: 46
  weak: 13, 15, 17, 46
  see also *localization*
Grassmannian: 46
group
  2-ary: 122
  classical: 47, 92
  commensurable: 114
  commutator: 47, 116
  definable: 110
  locally definable: 110
  nilpotent: 49
  orthogonal: 47
  settled: 122
  symplectic: 47
  see also *modular*

Harrington, Leo: 4
Higman's Lemma: 8, 63, 65
homogeneity: 3, 12, 96
  relative: 5, 7, 12, 42
homomorphism, affine: 112, 117

imaginary elements: 23, 170
  weak elimination: 7, 21, 23
independence: 7, 19
  2-independence: 85
  see also *rank; Theorem; type, amalgamation*
irreducible, algebraically: 154
isolation
  affine: 42

Jarden, Moshe: 11

Kantor, William: 2
Kantor/Liebeck/Macpherson ([KLM]):
  1, 2, 3, 4, 8, 13, 46, 47, 54, 97, 104, 149
Kim, Byunghan: 9, 10
Kinbote, Charles: not in the text
Knight, Julia: 170

Lachlan, Alistair: 1, 2, 5, 40, 104, 154, 155, 170
language, geometric: 143, 177
  see also *standard*
Lascar, Daniel: 9

Lie
  see *coordinatizable*
Liebeck, Martin: 2
linked
  see *orthogonality*
local definability
  see *group*
localization: 35

Macpherson, H. Dugald: 2, 47, 96
modular
  locally: 7, 101
  geometry: 24, 43
  structure: 7, 101
  see also *finite, basis*

nonmultidimensionality: 94, 120, 149, 150
nonstandard $\mathcal{M}^*$: 5, 6, 11, 12, 50, 177

orthogonality: 8, 31*ff*
  almost: 43
  joint: 52
  strict: 33
  see also *geometry*

Pillay, Anand: 10
Poizat, Bruno: 110
pre-rank: 18, 21
primitive: 2, 47
proper
  Grassmannian: 46
  model: 68
pseudofinite: 11
pseudoplane: 104, 154

quantifier
  dimension comparison: 6, 178*ff*
  $\exists^\infty$: 170
  elimination: 16, 20, 170
  Witt defect: 6
quasifinite
  axiomatization: 1, 8, 67, 75*ff*, 170
  structure: 5, 6, 11*ff*, 54, 56, 61

rank: 7, 18*ff*
  and measure: 7, 124*ff*
  of a geometry: 21
  see also *general position*
reducts: 4, 5, 6, 8, 61, 107*ff*, 141*ff*, 149*ff*
regular expansion: 40
rigid
  see *geometry*

section: 67*ff*
semiproper: 46
Shelah, Saharon: 2, 8, 9
Shelah degree: 88
shrinking: 1, 5, 182
    see also *envelope*
simple theory: 3, 8-10
skeleton: 67, 76
s_n^*: 5
Srour, Gabriel: 185
stabilizer: 110*ff*
standard
    language: 74
    ordering: 64
    system: 38
stretching
    see *envelope*
support: 70
    bounded part: 71
    reduced: 71

tame: 135
Theorem, Independence: 9
type
    amalgamation: 7, 9, 82*ff*, 103*ff*
    geometric: 77
    gtp: 122, 134
    of Grassmannian: 46
    semigeometric: 84

Wagner, Frank: 115
Witt: 44
    see also *defect; quantifier*
word: 64

Ziegler, Martin: 2, 63, 126
    finiteness conjecture: 79*ff*, 173
    see also *AZ enumeration*
Zilber, Boris: 1, 2, 3, 10, 40